Controlled Thermonuclear Fusion

Controlled Thermonuclear Fusion

Jean Louis Bobin
Université Pierre et Marie Curie (Paris 6), France

edp sciences

World Scientific

Published by

World Scientific Publishing Co. Pte. Ltd.

5 Toh Tuck Link, Singapore 596224

USA office: 27 Warren Street, Suite 401-402, Hackensack, NJ 07601

UK office: 57 Shelton Street, Covent Garden, London WC2H 9HE

British Library Cataloguing-in-Publication Data
A catalogue record for this book is available from the British Library.

Originally published in French as "**La Fusion Thermonucléaire Contrôlée**" by EDP Sciences.
Copyright © EDP Sciences 2011. A co-publication with EDP Sciences, 17, rue du Hoggar, Parc d'activités de Courtaboeuf BP 112, 91944 Les Ulis Cedex A, France.

This edition is distributed worldwide by World Scientific Publishing Co. Pte. Ltd., except France.

CONTROLLED THERMONUCLEAR FUSION

ISBN 978-981-4590-68-6

Typeset by Stallion Press
Email: enquiries@stallionpress.com

Acknowledgments

This book is a revised and updated version of a 2011 publication (in French) by EDP Sciences titled *Introduction à la fusion thermonucléaire contrôlée*. Several colleagues have helped in the making of the final manuscript in French. It is my pleasure to acknowledge discussions and comments by

- Daniel Heuer,
- Elisabeth Huffer,
- Jean Jacquinot, who wrote a preface to the French edition,
- Michel Le Bellac and Michèle Leduc, science editors at EDP Sciences,
- the late Gilbert Payan, who introduced me to the work of Robert W. Bussard, and
- Stephan Weber.

They all contributed many improvements to the initial draft.

Sophia Chen deserves special recognition for her careful reading of the manuscript in English and comments on points that needed clarification.

Contents

Foreword

Nuclear fusion powers the stars. Making this energy source work on Earth is an old challenge. In 1933, Lord Rutherford stated: *"Anyone who looks for a source of power in the transformation of the atom is talking moonshine."* It will be seen further down that indeed the beam target scheme Rutherford had in mind cannot be on the path towards an energy source. However, other ways that might succeed do exist. Controlled thermonuclear fusion would then deliver energy from an endless source — a dream for the future of mankind.

In order to realize the dream, scientists all over the world deliberately started inventing and developing new devices based on the results of fundamental research. Looking back at the history of technologies, we see examples that are to the contrary. On the one hand, the first industrial revolution is associated with the steam reciprocal engine, a technology that was developed empirically for centuries. The science (i.e. thermodynamics) came afterwards. It provided explanations and guiding principles for improvements. On the other hand, the science of electricity was developed first, followed by techniques in power supply and communication. In both cases, the development was a result of private investment and initiative. States played rather ancillary roles, granting concessions and permits.

During the 20th century, state-owned companies in mining, transportation, power supply, etc., were established for ideological reasons long after the occurrence of industrial revolutions.

The situation has been quite different in the realm of nuclear technologies. The birth of the nuclear industry took place during World War 2. After an initial stage devoted to nuclear weapons, research began on reactors best suited for power plants, first in national laboratories in which part of the activity is classified. Progressively after the war, principles and developments of civilian technologies were made public and also implemented thanks to private companies. However, after a rapid start, nuclear power plants developed slowly. Protests against nuclear power are not the only reason. So far, nuclear reactors have proved to be competitive only if the produced electric power exceeds 1 GW. In an economy without large-scale planning, the demand is lower and is best satisfied with fossil fuels that require locally smaller investments (Box 1). Only in France, a centralized country, a public operator (the state-owned *Electricité de France* (EDF)) was able to complete a program ending with electric generation being 80% nuclear. Actually, it was a public response to the first oil shock (1973). Governments are also expected to play a major role in a possible "nuclear renaissance".

Box 1. Nuclear gigantism

The larger the size of a single reactor, the more profitable is a fission power plant. As time elapsed, the rated electric power of the second-generation PWR (pressurized water reactor) increased from 500 MW to 1300 MW. In the third-generation EPR (European Pressurized Reactor), the rated power will be 1600 MW.

In order to build a nuclear reactor, a huge investment has to be allocated in a short time. Only communities that are large enough can afford it. Local demand is often at a smaller level, typically 500 MW, an amount coal and gas power plants can economically deliver if environmental issues are ignored.

In the history of energy technologies, many breakthroughs occurred after inventions or discoveries that were not primarily intended at creating energy sources. On the contrary, since the beginning of the 1950s, research on controlled fusion deliberately aim at future power plants. An international scientific community supported by the richest nations is at work. Scientists have the intuition that it will indeed be possible to eventually harness the energy of nuclear fusion. There are still open questions: When? How? What will be the cost?

Pioneering investigations started in the wake of the unusually fast development of nuclear (fission) energy. Most contributors worked in secrecy within the national laboratories built during or just after World War 2. In 1958, at the second Geneva conference "*Atoms for peace*", many results were declassified. Talks were given at dedicated sessions. Devices and posters from the U.S.A. were presented in a scientific exhibition. An impressive creativity dealing with magnetic confinement was thus disclosed. Although officially optimism prevailed about the near future of fusion, unexpected problems had just arisen. Indeed, in the early days of fusion research, plasma physics was still in infancy. Before massive studies were devoted to fusion plasmas, astrophysicists or theorist, among whom L. D. Landau is a prominent figure, had obtained some basic results. Known properties of widely investigated gas discharges (Langmuir) were too far from fusion conditions. A new chapter of physical science had to be written in order to accumulate knowledge about those states of matter that are of importance for fusion. The task is ongoing. For instance, although impressive advances were achieved, instability problems that appeared in the 1950s are still being investigated.

Breakthroughs occurred in the late 1960s. In Russia (then USSR), a device called "Tokamak", a Russian acronym for the toroidal chamber and magnetic fields, had far better performance (in terms of temperature and confinement time) than any other magnetic configuration tested so far. A team of British experimenters allowed to work at the Kurchatov institute validated the claimed results. It was the starting point for important, step-by-step developments. Today, a fourth-generation tokamak called the International Thermonuclear Experimental Reactor (ITER) is constructed at Cadarache, southern France, on behalf of an international consortium. It will be running in the 2020s.

Another breakthrough came about at the same time. The invention of lasers was followed by the experiments on interaction with matter. Laser-driven inertial confinement appeared as an alternative for fusion. Numerical simulations were encouraging, and dedicated high-power lasers were built, culminating with two megajoule devices: the National Ignition Facility (NIF) in Livermore, USA, and the Laser MegaJoule (LMJ) in France. First attempts at the ignition of a small mass of a deuterium–tritium mixture were performed at the NIF in 2012.

Since the 1970s, the history of fusion research is the history of two main programs: tokamaks and laser-driven inertial fusion. In both cases, significant results are expected to appear during the coming decades. The odds seem in favour of scientific success, but whether they will open the way towards the development of economically viable energy sources is still an open question.

In thermonuclear fusion, several branches of physical sciences are involved: some nuclear physics, a lot of atomic physics that will not be dealt with in this book, and a huge amount of plasma physics. A large part of the following is devoted to the physical bases. The presentation is quite general in the first three chapters. More details are given in Chapter 4 and 7. Introduction to the main programs is dealt with in Chapter 5 (tokamaks), 6 (ITER) and 8 (megajoule lasers and other big drivers). In Chapter 9, some off-the-main-trail proposals are presented. Are they science fiction? No more presumably than tokamaks or laser fusion since so far no positive energy balance was achieved in any fusion device. Chapter 10 is devoted to nuclear fusion reactors as they can be imagined in 2013.

Units

This book is about energy, a quantity whose unit in the SI system is the joule (J), i.e., the work done by a 1 meter displacement in the direction of a 1 newton force. The watt (W) (1 watt $= 1$ joule/second) is the unit power. In real life, the following practical units are commonly used:

1 kilojoule (kJ) $= 10^3$ J
1 megajoule (MJ) $= 10^6$ J
1 gigajoule (GJ) $= 10^9$ J
1 terajoule (TJ) $= 10^{12}$ J
1 kilowatt (kW) $= 10^3$ W
1 megawatt (MW) $= 10^6$ W
1 gigawatt (GW) $= 10^9$ W
1 terawatt (TW) $= 10^{12}$ W

In atomic, nuclear and particle physics, the following specific units are used:

1 electron-volt (eV) $= 1.6 \times 10^{-19}$ J
1 kilo-electron-volt (keV) $= 1.6 \times 10^{-16}$ J
1 mega-electron-volt (MeV) $= 1.6 \times 10^{-13}$ J
1 giga-electron-volt (GeV) $= 1.6 \times 10^{-10}$ J
1 tera-electron-volt (TeV) $= 1.6 \times 10^{-7}$ J

Energy and Temperature

The temperature of a gas at thermal equilibrium is proportional to the mean kinetic energy of its microscopic particles in random motion. The proportionality coefficient is the Boltzmann constant $k = 1.38 \times 10^{-23}$ J/K. The electron-volt, the energy unit of nuclear physicists, is related to the Kelvin, the usual temperature unit, according to the rule of thumb:

$$1\,\text{eV} \Leftrightarrow 10^4\,\text{K}.$$

Therefore, high temperatures on the human scale correspond to very low energies on the subatomic scale. In the following, quite often, temperatures are given in energy units.

1

Some Basic Physics

1. Atoms

In the early years of the 20th century, the atomic structure as it is known today was unravelled through a famous experiment, in which alpha particles from a radioactive source were impinging onto metal foils and unexpectedly, some of the particles were backscattered. As Rutherford put it later: "*it was as if you fired a 15 inch shell at a piece of tissue paper and it came back and hit you.*" The effect can be readily explained by thinking of alpha particles as microscopic particles each carrying a positive electric charge. Each of them undergoes electrostatic repulsion from a small-sized nucleus, positively charged, sitting at the centre of the atom (Figure 1.1). This is now known as Rutherford scattering, which is actually a special case of *Coulomb scattering* (Box 1.1). Meanwhile, alpha particles were identified as helium nuclei.

An atom is made of a central nucleus whose positive charge is exactly compensated by the negative charge of the surrounding electron cloud. Now, the electric charge is quantized. The elementary charge, i.e. the smallest quantity that can be isolated, is denoted by $e = 1.6 \times 10^{-19}$ coulomb. An electron carries a charge of $-e$. The charge of a nucleus is an integer Z times e. Every *atomic number* Z corresponds to a given chemical element.

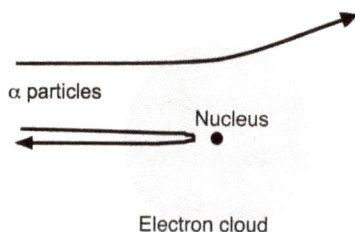

α particles

Nucleus

Electron cloud

Figure 1.1. The scattering of helium nuclei (α particles) by an atom. In the electron cloud, negative charges are smeared within a 10^{-10} m radius sphere. Their influence on incoming charged particles is negligible. On the contrary, in the heavy nucleus, the electric charge is concentrated in a sphere of radius 10^{-15} m. In most collisions, the projectile remains far from the target and is hardly deviated. Whenever it comes closer to the nucleus, it is backscattered.

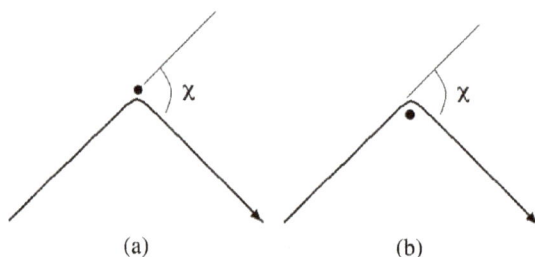

χ

χ

(a)   (b)

Figure 1.2. Coulomb collisions in the target's frame: a) both particles have the same sign, repulsive force; b) particles have opposite sign, attractive force. χ is the scattering angle depending upon the direction and the energy of the projectile.

Box 1.1. Coulomb scattering

Rutherford scattering is a special case of encounters between charged particles. The dynamics of the process are driven by the Coulomb electrostatic force, which follows the inverse square law, hence the name Coulomb collision. In the targets frame, trajectories follow branches of hyperbolas. They avoid the centre of force when the charges have identical signs, and they turn around it when the charges have opposite signs.

A familiar albeit misleading picture (Figure 1.3) shows the electron cloud as objects orbiting the nucleus the same way as planets orbiting the Sun. Actually, the atom is a complex structure obeying the laws of quantum physics. There are no such things as sharply defined trajectories. Electrons occupy quantum states corresponding to *energy levels*. Whenever a transition occurs from a given energy level to another one, a photon

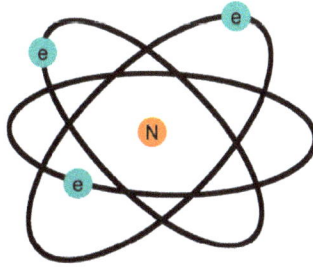

Figure 1.3. Misleading planetary model of the atom: electrons (green dots) orbit the nucleus (orange dot). The figure is not drawn to scale: the radius of the nucleus is about 1/100000 of the atomic radius.

is emitted or absorbed whose frequency ν is proportional to the energy difference ΔE between the two level according to the well-known Planck's formula:

$$\Delta E = h\nu$$

where h ($=6.63 \times 10^{-34}$ Js) is a universal constant.

An atom having lost at least one electron is *ionised*.

2. Nuclei

The nucleus is an assembly of nucleons: protons carrying the positive elementary charge e, and neutrons, which are electrically neutral as their name says it. The number of protons is equal to the atomic number Z. A proton and a neutron are both 2000 times more massive than the electron. Consequently, the greater part of the atomic mass lies in the nucleus. Since the radius of the nucleus is 10^5 times smaller than the atomic radius, the density of the nuclear matter is 10^{15} times the solid-state density.

The zoo of nuclei is very rich. Each one is characterized by two numbers: the number of protons Z and the total number of nucleons A, which is also the atomic mass number. Given a chemical element, several mass numbers are available for a single atomic number. The corresponding nuclei are called *isotopes*. A nucleus is represented in short hand by a symbol with two indices on the left: the lower index is the atomic number Z, the upper one is the mass number A. Accordingly the deuteron, a heavy isotope

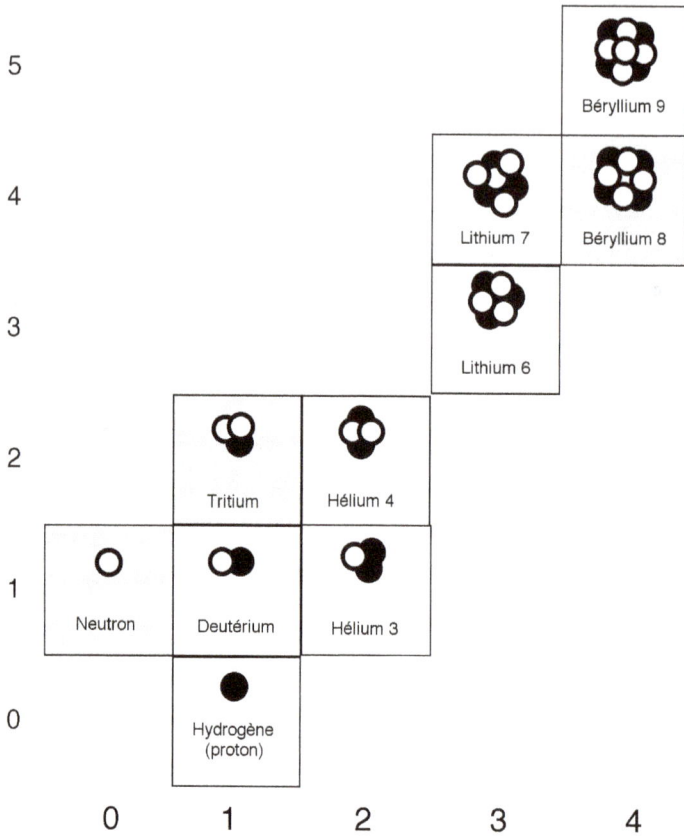

Figure 1.4. Light nuclei. In abscissas the number of protons, in ordinates the number of neutrons. Gray cells correspond to unstable isotopes. Beryllium-8 is by far the most unstable of light isotopes.

of hydrogen, is represented by $^{2}_{1}$H. Many isotopes are stable: most isotopes are unstable and decay into smaller objects, according to various modes of *radioactivity*. On Figure 1.4, light nuclei are displayed on a Segré's chart.

If only electrostatic repulsion and gravitational attraction, a small, almost negligible force, existed, a nucleus made of protons and neutrons would explode. A compensating force of a different nature is necessary to bind an assembly of nucleons together. It should be exceedingly intense and is named the *strong nuclear interaction*. Its range cannot extend much beyond the radius of the nucleus. Otherwise, Rutherford scattering of alpha particles by nuclei would not obey Coulomb's law, as observed experimentally, whenever the incident particle comes closer to the target.

The electromagnetic interaction with infinite range is mediated by a massless particle, i.e. the photon. Due to its short range, the strong nuclear interaction is, on the contrary, mediated by massive particles: pions which are created and used in high energy physics experiments. They are either neutral or charged (either sign). Their mass is about 200 times the electron mass, hence the name meson which stands for an intermediate between electrons and nucleons. Nucleons and pions are themselves made of smaller entities — quarks, which are quoted here for the sake of completeness. Actually, an assembly of nucleons stuck by pions is the most adequate picture in the realm of man-made nuclear energy.

In nuclear processes such as the neutron decay (β radioactivity)

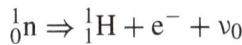

$$_0^1 n \Rightarrow {}_1^1 H + e^- + \nu_0$$

electrons and neutrinos (ν_0) are involved. Since these particles are insensitive to the strong interaction, the above nuclear reaction cannot depend on the strong force. Another nuclear force is necessary. It turns out to be far less intense and for that reason is called the *weak interaction*.

As far as we know it, only four forces do exist in nature: gravitation and electromagnetic interaction which have infinite range and the two nuclear interactions whose range is comparable to the size of a nucleus. Inside many nuclei, the strong nuclear force is more intense than the electrostatic repulsion. Nucleons stay confined inside a *potential well*. On the contrary, when Z is larger than 82 (Pb), the global electrostatic field is a sizable fraction of the strong attractive field. Heavy nuclei are thus unstable (radioactive).

3. Nuclear reactions

In some chemical reactions, atoms combine to build up more or less complex molecules. In other chemical reactions, molecules exchange atoms or group of atoms. Such processes absorb or release energy. By the same token, nuclei interact and exchange nucleons whilst absorbing or releasing energy. For instance, an alpha particle striking a nitrogen nucleus transforms the latter into an oxygen nucleus and a proton is emitted. Such a transmutation is written the same way as a chemical reaction. In a nuclear reaction, the number of nucleons and the total electric charge are conserved.

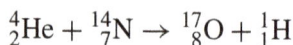

$$_2^4\text{He} + {}_7^{14}\text{N} \rightarrow {}_8^{17}\text{O} + {}_1^1\text{H}$$

The probability of the reaction is represented by a quantity whose dimensions are those of a surface: the *cross section*. This definition applies to any binary processes.

Box 1.2. The probability of a microscopic event and cross section

Most nuclear reactions are binary since they have two reactants. Every event has a probability of occurrence. Let J_p be the flux of particles in a parallel beam, all with the same velocity. The beam impinges normally onto a slice of matter with thickness dx containing n_c targets per unit volume. Provided an incident particle undergoes no more than a single collision to produce a given effect, the variation of the flux through the slice is proportional to the flux itself and to the density of targets, viz. $\frac{dJ_p}{dx} = \sigma n_c J_p$, in which σ has the dimension of a surface. σ is the *cross section* for the microscopic event. It is an indirect measure of its probability. It depends on the projectile velocity (energy) relative to the target.

A practical unit well suited to nuclear reactions is the *barn*, i.e. 10^{-28} m^2.

When nucleons are put together in a stable nucleus, the resulting mass is smaller than the total of the individual masses. The *mass defect* evidenced in mass spectrometry is a few thousandths of the total. However, in energy units, thanks to $E = mc^2$, the mass defect is attractively large. Now, the mass defect per nucleon as a function of the atomic mass A is presented in Figure 1.5. A flat maximum is obtained for chemical elements in the middle of the periodic table.

Energy from nuclear reactions is expected in two cases:

(1) breaking heavy nuclei, as in the fission of uranium. Since the mass defect per nucleon is larger in the middle of the chart, the total mass of the fission products is smaller than the initial mass. The lost mass is converted into energy. Such a process occurred two billions years ago in the natural reactors at Oklo (Gabon).

(2) combining light nuclei to form a heavier one. In such a fusion process, the mass of the created nucleus is smaller than the total mass of the

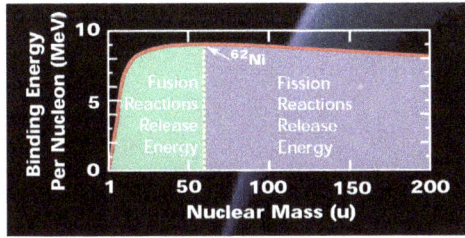

Figure 1.5. The binding energy of nuclei (mass defect) per nucleon is maximum for elements in the middle of the periodic table for mass numbers (nuclear mass in atomic units) ranging from 30 to 80. Consequently, fusion reactions between light nuclei as well as fission of heavy nuclei release energy (CPEP document [1]).

reacting nuclei. Energy is released. Fusion reactions (starting with hydrogen) power the Sun and other stars. On Earth, deuterium (Box 1.3) is more appropriate for an energy source.

Fission has been used for a long time in nuclear power plants. However, controlled fusion has not yet been demonstrated. Important research programs have been active for the last 60 years. Results apparently come at a slow pace. However, real advances have been achieved in an undertaking much more difficult than initially anticipated.

Box 1.3. The deuteron as a singular nucleus

Deuterium is a heavy isotope of hydrogen. Its nucleus, the deuteron is made of a proton and a neutron. It has a unique property: it is stable although loosely bound. Its centre of mass is different from its electrostatic centre of force. A collision with another nucleus is more like the impact of two projectiles. Fusion reactions with deuterons have the largest cross sections of all fusion reactions. They are also more easily obtained in the laboratory.

Reference

1. http://fusedweb.llnl.gov/CPEP/Images/CPEP-Fusion-2000-EN-Front.PDF.

2

Thermonuclear Reactions

1. Fusion reactions

A fusion reaction is a subatomic process in which two nuclei unite in order to form another nucleus with more nucleons. Fusion begins with the simplest nuclei, i.e. protons. Step by step, all elements of the periodic chart can be created in this way. The early Universe and active stars are element-making factories.

Binding energies of light elements are presented in Figure 2.1. Fusion reactions involving protons (H, mass number 1) or heavier isotopes of hydrogen (deuterium D, mass number 2, and tritium T, mass number 3) produce Helium-4. According to the discussion in the previous chapter, such reactions release a large amount of energy.

On the microscopic scale, a fusion reaction proceeds through several stages. The first stage is the collision of two nuclei. Since both nuclei are positively charged, repulsive electrostatic forces must be overcome.[1] The target appears protected by a kind of palisade, the electrostatic potential barrier surrounding a potential well (Figure 2.2) whose radius is determined by the competition with the nuclear interaction that might be involved. The barrier is higher in the case of a weak interaction.

---

[1] This problem is inexistent in neutron-induced fission.

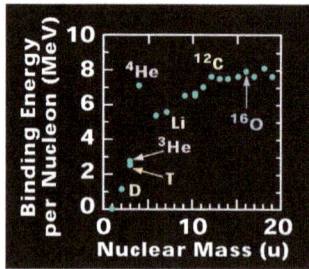

Figure 2.1. The binding energies of light nuclei. Creating a Helium-4 nucleus with heavy hydrogen isotopes (D and T) release large amounts of energy (CPEP document).

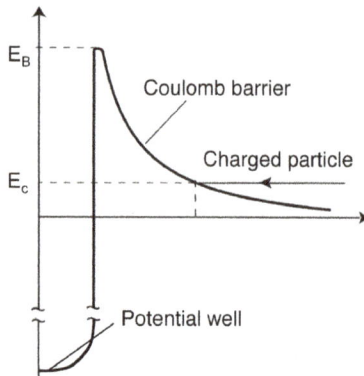

Figure 2.2. Electrostatic potential barrier (Coulomb force) around a nucleus. An incident ion with kinetic energy E_c smaller than the height E_B of the barrier needs tunnelling effect (see text) to access the potential well.

Having overcome the potential barrier, the light nuclei unite to create an unstable *compound nucleus*. This second stage is immediately followed by the third: the decay of the compound into a nucleus heavier than each of the initial components plus the by-products. If the latter are electrically charged, they escape through the potential barrier (the fourth and final stage).

This series of processes illustrates the importance of potential barriers in nuclear fusion reactions. Crossing a barrier can be made in two ways: jumping over or getting through it. In classical mechanics, only the former is permitted. The energy of the projectile should be higher than the barrier, a few MeV in the case of the strong interaction between hydrogen isotopes. As it will be shown later, the energy balance would be unfavourable.

However, at subatomic distances, quantum physics prevails. Particles are endowed with de Broglie waves which look like optical waves. These are not completely reflected by obstacles. Part of the amplitude goes through. Whenever the energy of the incident particle is smaller than the barrier height, there is a probability, which can be calculated, that it will tunnel through the barrier and reach the opposite side. This holds for stages 1 and 4 of the fusion process.

At low energy, cross section of fusion reactions depends on the barrier transparency, which can be calculated in quantum mechanics, given the incident particle energy. Such calculations are mandatory since in many cases the cross sections are too small to be measured experimentally.

Jumping over the potential barrier or tunnelling through it requires energy. The problem of energy balance arises at the microscopic level.

2. Hydrogen — the fuel of the Sun and other stars

In Sun-like stars, four hydrogen nuclei build up a Helium-4 nucleus through the Carl von Weizsäcker's sequence of reactions (Table 2.1). In other types of stars, the same result is catalyzed by carbon and nitrogen nuclei (Bethe's cycle).

The first reaction in the sequence is known as a *proton–proton* reaction. It is a weak interaction whose products are a deuterium nucleus, a positron and a neutrino. Its probability of occurrence is so low that until now it has been impossible to observe it in the laboratory. Current instrumentation is not sensitive enough, thus no cross section has been measured. However, the reaction does exist in the centre of the Sun as it is known from calculation and modelling results that are consistent with the observed behaviour of the star, or from neutrinos that hit Earth. Some of these particles are caught

Table 2.1. The Weizsäcker's sequence.

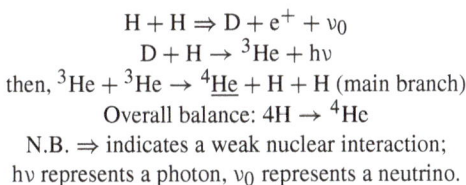

$$H + H \Rightarrow D + e^+ + \nu_0$$
$$D + H \rightarrow {}^3He + h\nu$$
then, ${}^3He + {}^3He \rightarrow {}^4\underline{He} + H + H$ (main branch)
Overall balance: $4H \rightarrow {}^4He$
N.B. $\Rightarrow$ indicates a weak nuclear interaction;
$h\nu$ represents a photon, ν_0 represents a neutrino.

in large detectors whose main component is a huge mass of water (a few hundreds to a few thousands tonnes). Unwanted background noise from cosmic radiation is avoided by building neutrino detectors under a few kilometres of rocks in mines no longer in use or in mountain-crossing tunnels. Thus, emitted neutrinos make the centre of the Sun "visible".

In the conditions presumably prevailing in the centre of the Sun (temperature of about 1.5×10^7 K, particle density of 10^{32} per m^3), half the hydrogen is to be consumed in some 7 billions years.

The exceedingly slow proton–proton reaction drives the dynamics of the entire sequence for the benefit of a long-lived solar system. As a counterpart, such reactions would be ineffective as a terrestrial energy source. On the contrary, provided they are harnessed, nuclear reactions with heavy hydrogen isotopes would be suited to the job.

3. Deuterium reactions

In the Sun, a deuteron reacts with a proton (the strong interaction) and thus disappears as soon as it is created. The deuterium nucleus also reacts with another deuterium nucleus giving birth either to tritium, the superheavy unstable hydrogen isotope, or Helium-3. The deuterium nucleus also reacts with either one of these two new nuclei (Table 2.2).

Deuterium reactions are strongly exothermic, especially for D–T and D–^{3}He reactions. Below 50 keV, fusion cross sections are given by the Gamow's formula:

$$\sigma = \frac{S}{E} e^{-\frac{G}{\sqrt{E}}}$$

where the S/E factor comes from the kinematics of the reactive collisions whilst the exponential term is the transparency of the potential barrier

Table 2.2. Deuterium reactions.

$$D + D \rightarrow T \ (1.0\,\text{MeV}) + H \ (3.0\,\text{MeV})$$
$$D + D \rightarrow {}^3\text{He} \ (0.8\,\text{MeV}) + n \ (2.5\,\text{MeV})$$
$$D + T \rightarrow {}^4\text{He} \ (3.5\,\text{MeV}) + n \ (14.1\,\text{MeV})$$
$$D + {}^3\text{He} \rightarrow {}^4\text{He} \ (3.7\,\text{MeV}) + H \ (14.7\,\text{MeV})$$

Table 2.3. The coefficients of Gamow formulas for the reactions of Table 2.1.

Reaction	S (barn · keV)	G ($\sqrt{keV}$)
$D + T \rightarrow {}^4He + n$	11000	34.4
$D + D \rightarrow T + H$	53	31.4
$D + D \rightarrow {}^3He + n$	53	31.4
$D + {}^3He \rightarrow {}^4He + H$	7000	88.8

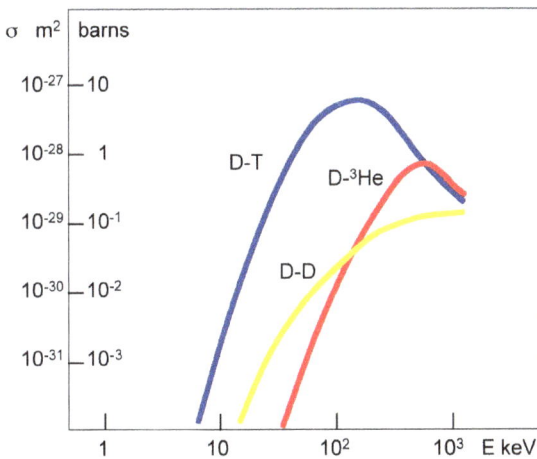

Figure 2.3. Cross sections as functions of the projectile energy for deuterium reactions. Measurements are usually performed by sending deuterons onto gaseous tritiated targets [2]. Below 25 keV, measuring devices are not sensitive enough. Cross sections are then evaluated by calculation only.

calculated after the tunnelling effect. Numerical coefficients S and G are specific to each reaction (Table 2.3).

Cross sections were determined experimentally for energies greater than 25 keV and by calculation only at lower energies [1]. They are displayed in Figure 2.3.

The cross section of the deuterium–tritium reaction (D–T) has a maximum for an energy of about 100 keV. This peak is due to a resonance effect (excited state of the compound nucleus) in the nuclear process. By the same token, the reaction D–^{3}He has also a resonance peak at 500 keV instead of 100 keV. Cross sections are significantly smaller and compare with D–D reaction cross sections.

Each one of D–T and D–^{3}He reactions releases an amount of energy much larger than the total obtained from both D–D reactions. At low energy, the D–T reaction is the most likely. Although its cross section is small according to nuclear physics standards, applications are foreseeable at an acceptable cost of energy. Only the D–T reaction has been considered in research projects conducted in the last 50 years in order to end up with a controlled energy generation.

However, tritium is a radioactive isotope with a half-life of 12 years. Except as transient traces, it does not exist in nature. It has to be manufactured using neutron-induced lithium fission reactions:

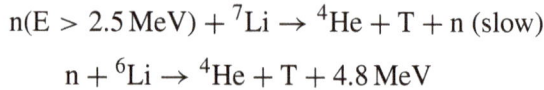

$$n(E > 2.5\,\text{MeV}) + {}^7\text{Li} \rightarrow {}^4\text{He} + T + n \,\text{(slow)}$$

$$n + {}^6\text{Li} \rightarrow {}^4\text{He} + T + 4.8\,\text{MeV}$$

whose cross sections are given in Figure 2.4 [1].

Below 100 keV, fusion cross sections increase rapidly with energy. For energy amplification, staying in the keV range is advisable since the

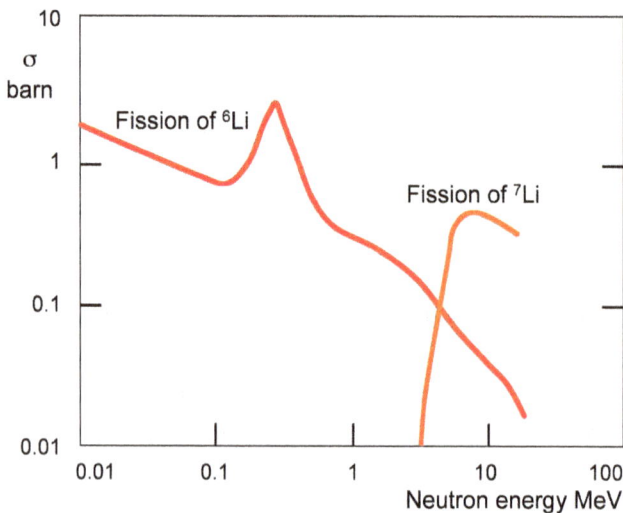

Figure 2.4. Tritium generation cross sections from lithium as functions of neutron energy. The low-energy cross section for Lithium-6 fission is comparatively large. The threshold for Lithium-7 fission is about 2.5 MeV.

energy released by fusion is in the MeV range. Accelerated nuclei impinging onto a target has been proven ineffective. Indeed, they encounter other charged objects, ions and electrons. In the same way as Rutherford's alpha particles, they undergo numerous unavoidable Coulomb collisions (see Box 1.1). They slow down before entering an unlikely nuclear reaction process. Stars are a counter example. Fusion reactions take place within a hot ionized gas (plasma) and thus generate energy. Such conditions are worth a closer look.

4. The thermonuclear regime

Inside an ionized gas, ions move freely from one collision to the next one. In thermal equilibrium, the distribution of their velocities (hence of their energies) is larger and shifted towards higher energies as temperature increases. The total number of fusion reactions per unit volume per second is proportional to the particle densities of the reacting species and to a quantity denoted by ⟨σv⟩ usually called the *reaction rate* or *reactivity*. This temperature-dependant parameter is specific to a given reaction. It reads

$$\langle \sigma v \rangle = \int_0^\infty \sigma(E)\, n(E)\, dE$$

where E is the kinetic energy of the particles in an isotropic medium (see Box 2.1).

In a Maxwell–Boltzmann equilibrium, the high-energy tail of the distribution function n(E) is close to the exponential $\exp(-E/T)$ where T is the temperature in energy units. Such a function is rapidly decreasing for kinetic energies E well above the average represented by the temperature. On the contrary, in the same range, the cross section grows at a very high rate. As shown in Figure 2.5, the product of the distribution function and the cross section has significant values, a peak, only near an energy E_M [3].

The kinetic energy of those nuclei which contribute most to fusion is thus much higher than the thermal energy — typically 6 to 10 times. A narrow slice of the distribution is depleted by the reaction. It is constantly replenished since the gaseous medium is in thermodynamical equilibrium.

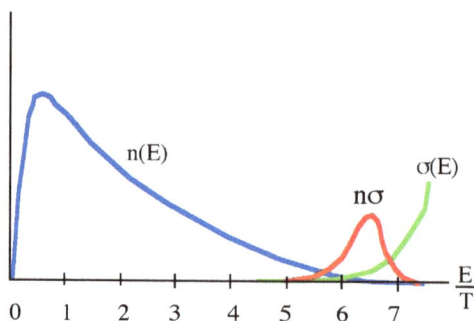

Figure 2.5. The product $n\sigma$ of the cross section $\sigma(E)$ and the energy distribution $n(E)$ at a given temperature T is significant only in the vicinity of an E/T value between 6 and 10.

Isotropy in a high-temperature plasma is a further advantage of the thermonuclear regime with respect to a beam-target geometry (Box 2.1). The whole volume participates in the reaction. Consider 100-keV D^+ ions (maximum cross section for the D–T reaction is $5 \times 10^{-28}\,\text{m}^2$). Their velocity is 10^6 m/s. A 0.1-ampere (10 kW) beam of such ions with cross-sectional area 1 mm^2 impinging onto a 1-mm gas jet with density one tenth atmospheric would produce 3×10^{12} reactions per second, i.e. 10 W fusion power. A 10^{20}-ions/m^3 plasma with temperature 10 keV produces every second 4×10^{18} reactions per m^3, i.e. 10 MW/m^3. Isotropy is the key to large volumes of matter participating in the reactions.

Box 2.1. The thermonuclear regime and the importance of isotropy

In accelerator physics, *luminosity* is the quantity L whose product with the cross section is the number of reactions per unit time $\frac{dN_R}{dt} = L\sigma$. In a beam-target geometry, luminosity is the product $L = n_F v n_C \lambda_{eff} \Sigma$ where n_F is the particle density in the ion beam, v is their velocity, n_C is the particle density of the target, λ_{eff} is the effective interaction thickness and Σ is the beam cross-sectional area. In the thermonuclear regime, the velocity v_m which maximizes $n\sigma$ is attributed to the projectiles. Then, $L = n_F v_m N_C$ where $N_C = n_C V \gg n_C \lambda_{eff} \Sigma$ is the number of targets inside the entire plasma volume V.

Reaction rates for deuterium are displayed in Figure 2.6. In the case of the D–T reaction, the rates have comparatively high values when the temperature is above 10 keV (100 millions degrees Celsius).

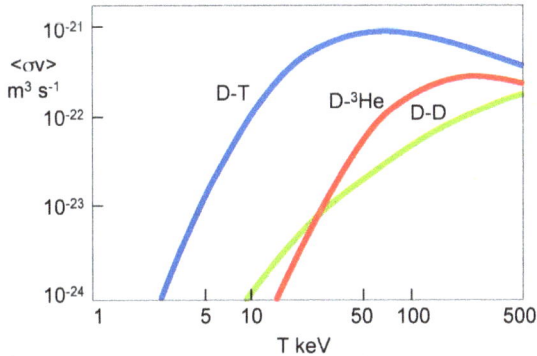

Figure 2.6. Reaction rates for D–T, D–^{3}He and D–D (total of both branches) in the thermonuclear regime. Thanks to the peak at 100 keV, the D–T reaction rate turns out to be high whenever the temperature of the reacting medium is a few tens of keV.

The rates of deuterium reactions are regularly updated: measurements improve with time when theoretical advances are combined with the use of powerful computers.

The thermonuclear reaction is a remarkable *energy amplifier*. Of course, energy have to be invested in order to give ions and the accompanying electrons an average energy corresponding to the temperature. On the production side, the energy released by the fusion reaction is tens to hundreds times higher. The reciprocal of this ratio is the order of magnitude of the minimum fraction of the fuel which should react in order to achieve a positive energy balance.

5. The ignition and sustainability of a thermonuclear reaction

100 million degrees Celsius is a temperature beyond terrestrial norms. The fusion challenge is heating plasma up to such a temperature in conditions that should last long enough for a positive energy balance.

In nature, thermonuclear reactors, i.e. stars, are born from gaseous clouds in interstellar space. When the mass of the cloud exceeds a threshold value, the cloud is unstable (Jeans criterion). Due to gravitational attraction, collapse begins, at first slowly. A motion is set up towards an accumulation point. Nucleation of a denser medium takes place with an enhanced attractive power. Consequently, the process accelerates. Since there is practically no heat exchange with the outer universe, the gas is compressed adiabatically. Heating occurs, more so in the vicinity of the centre. Furthermore, at the centre, kinetic energy is converted into thermal energy. These effects contribute to the creation of a hot, dense plasma ($T > 10^6 \, K \approx 100 \, eV$). Thermonuclear reactions start providing further heating. The subsequent pressure gradient slows down and eventually stops the compression. A new equilibrium is established. Stellar life begins. A steady state prevails in which energy gains from the nuclear reaction exactly compensates radiation losses.

On Earth, various ways of plasma heating are available. They are investigated in further chapters. At first, it will be shown that thermonuclear reactions can be made self-sustainable.

Consider, for instance, the D–T reaction. It produces a neutron and a helium nucleus. In most practical cases, the neutron escapes. Stopping it would require a very large thickness of hydrogen. Consequently, the external detection of neutron is a good diagnostic tool for the occurrence of fusion reactions. On the contrary, the helium nucleus, electrically charged, undergoes Coulomb collisions. It slows down inside the plasma and contributes to the heating (internal energy gain). The situation is about the same for both D–D reactions. Neutrons escape; charged reaction products contribute to heating.

In all cases, energy gains are due to charged reaction products only. The specific heating power is proportional to their energy, to the square of the ion density and to the reaction rate whose temperature dependence is thus of paramount importance.

Many mechanisms contribute to energy losses: contact with a wall, escaping particles and radiation. Losses can occur from the whole volume or through the surface. Radiative volume losses are essentially *bremsstrahlung* (Box 2.2), a sequel of electron–ion Coulomb collisions. In the case of fusion plasmas, bremsstrahlung losses are an unavoidable hindrance.

Most plasmas contain impurities, i.e. ions other than hydrogen isotope nuclei. These high-Z emitters contribute to further radiative losses through enhanced bremsstrahlung and line emissions.

In a fusion plasma, electron–ion collisions generate radiation. This is a general property of accelerated charges, e.g. when they move in the electrostatic field of another charge. In a Coulomb collision with an ion, an electron is actually losing energy, hence the name *bremsstrahlung* (German: literally, radiation from braking). This is an unavoidable energy loss. In the case of an optically thin plasma, photons created by bremsstrahlung escape. The radiated power due to this volume emission is proportional to the square root of the temperature and to the square of the particle density.

Energy gain from fusion and radiative losses by bremsstrahlung have the same dependence upon plasma density. Temperature ranges for which gains are greater than losses are readily determined on a diagram [4] showing gains and losses per unit volume as functions of T (Figure 2.7). Thresholds are evidenced for the self-sustainability of fusion reactions.

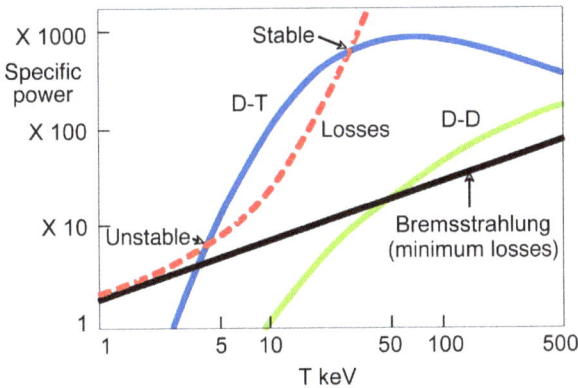

Figure 2.7. Looking for self-sustainability conditions in the cases of reactions D–D (both branches, green curve) and D–T (blue curve). Gains by energy deposition from charged reaction products overtake bremsstrahlung losses (black line) when the temperature surpasses a threshold specific of each reaction. The threshold itself is not a stable equilibrium point. Indeed, when the temperature increases beyond the threshold, gains are greater than losses, and the medium heats up without limits (thermal instability). In real life, losses are superior to mere bremsstrahlung (broken red curve). Gain and loss curves cross again at a point which represents a stable equilibrium. Indeed, should the temperature rise, losses are greater than gain and cooling of the medium towards equilibrium occurs. Conversely, in the case of a temperature decrease, gains superior to losses drive the medium back to equilibrium. The fusion reaction is then controllable.

According to more precise calculation,

D–T	temperature should exceed 4.4 keV
D–D (2 branches)	temperature should exceed 48 keV

Box 2.3. Burn efficiency

In a magnetically confined plasma, the projectile will always react (100% efficiency) if it stays confined for a time τ_R equal to the mean free path divided by the thermal velocity. Accounting for the velocity distribution, $\tau_R = \frac{1}{n\langle\sigma v\rangle}$. The burn efficiency η_C has the same order of magnitude as the ratio of the confinement time τ_C (usually given by experiments) to τ_R. It turns out to be equal to the product of temperature dependent $\langle\sigma v\rangle$ by $n\tau_C$, viz. $\eta_C \approx \frac{\tau_C}{\tau_R} = \langle\sigma v\rangle n\tau_C$.

In the inertial confinement regime, a hot plasma sphere of radius r is short-lived. Then, the burn efficiency is the ratio of r to the mean free path, viz. $\eta_c \approx \frac{r}{\lambda} = \sigma n r$, where the important parameter is the product nr of the radius and the particle density. In practice, the chosen relevant parameter is the areal density ρr where ρ is the mass density.

Looking at Figure 2.7, the control of fusion reactions can conceptually be achieved by following either one of the following two paths:

- On the one hand, assuming losses depend on temperature in such a way that a stable equilibrium point does exist. Among those losses, cooling by contact with a wall is excluded since it would immediately quench the reaction. In order to maintain fusion conditions as long as possible, the plasma can be separated from any kind of wall with a magnetic field. Such a *magnetic confinement* implies a plasma density smaller than 10^{20} ions per m³ in order to limit energy densities at a level compatible with the strength of the structural materials. The key point for a steady regime is the control of losses. In such optically thin plasmas, radiative losses are strongly dependent on impurities. The content of ions with charges higher than one is to be carefully tuned. Other losses stem from leaking particles through the outer surface of the plasma. Minimizing such losses implies a higher volume-to-surface ratio i.e. increasing the size of any given geometry. This is among the reasons why advances in magnetic confinement were achieved with bigger and bigger machines.
- On the other hand, giving up the search for a stable equilibrium, the system would be free to run away and burn a large amount of its fuel

content. The total energy released at the end of the process should exceed the energy invested to achieve thermonuclear burn. This lack of confinement is called *inertial confinement* [5], a major field of application for high-power lasers (Chapter 7 and 8).

At 100 million K, each nucleus entering the D–T reaction has an average energy of 10 keV, which is also the energy of the neutralizing electrons. The total input is 40 keV, compared to the 17.6 MeV of the reaction products. The amplification factor is 440. Consequently, a low burn fraction (>0.25%) might be enough for a positive energy balance in a steady-state reactor whose fuel is slowly renewed. On the contrary, an efficient explosive regime implies a high burn fraction (>30%).

6. The steady regime: the Lawson criterion

John Lawson's derivation is typical of an engineer in charge of designing an energy source. Such a device is expected to work according to the energy cycle sketched in Figure 2.8.

The thermonuclear burn efficiency is proportional to the product $n\tau_C$ of the particle density in the plasma and the confinement time (see Box 2.3). Consider the loop in Figure 2.8. The device is to produce more energy than invested in the whole process if, for a given efficiency η_T of the thermal

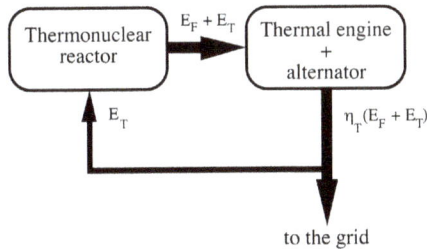

Figure 2.8. Energy cycle for an electric-power-producing magnetically confined reactor. The energy output after the plasma confinement time τ_C is the energy E_F released by fusion to which the thermal energy E_T of the plasma should be added. A thermal engine with efficiency η_T drives an alternator which delivers energy to the grid, part of it, i.e. E_T being used to heat the plasma for the next cycle. A positive balance is obtained whenever the energy available from the alternator is larger than E_T.

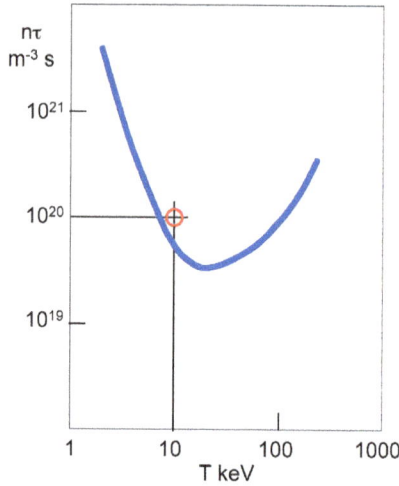

Figure 2.9. Looking for a criterion of the steady regime. The energy balance is positive if the point representing the plasma and confinement conditions lies above the curve plotted after Lawson. Conventionally, the conditions to be aimed at in magnetic fusion are at the centre of the red circle: 10 keV and 10^{20} m^{-3} s.

engine, the product $n\tau_C$ is greater than a certain function of the temperature. This function is plotted in Figure 2.9 for the D–T reaction assuming the canonical value 0.3 for the thermal engine efficiency. It shows a minimum of about 4×10^{13} cm^{-3} s (4×10^{19} m^{-3} s) for a temperature slightly above 10 keV.

For the sake of simplicity, in order to define conditions that a magnetic confinement reactor is expected to fulfill, the following easy-to-memorize criterions are chosen:

$$n\tau_C > 10^{20} \, \text{s} \, \text{m}^{-3} \quad \text{at} \quad T = 10 \, \text{keV}.$$

At temperature 10 keV, a particle density of about 10^{20} m^{-3} is an upper limit, given the confining magnetic fields that are attainable in volumes as large as hundreds of cubic meters. With this density, the confinement time is to exceed 1 second in order to fulfill the Lawson criterion. Simple figures to remember.

7. The explosive regime

In inertial confinement, the approach is completely different. Since there are no steady conditions, the reaction should release as much energy as possible before the system blows off. Consider a bare sphere made of a deuterium–tritium mixture at a temperature of at least 10 keV. This fluid medium is set into motion due to a rarefaction wave whose head propagates from the surface towards the centre (Figure 2.10).

Although over simplified, the scheme proves suited to investigate the main aspects of inertial fusion. Let f be the fraction of the initial content of tritium that participated in the reaction. This measure of the burn efficiency depends upon the areal density ρr of the compressed sphere (see Box 2.3) and through $\langle \sigma v \rangle$ upon the temperature T. In Figure 2.11, the variation of f as functions of ρr are displayed for different values of T. A burn efficiency superior to 10% is obtained whenever ρr is larger than 1 gcm^{-2}.

In order to obtain a positive energy balance in inertial fusion, the product fQ of the tritium burn fraction and the energy Q released in a

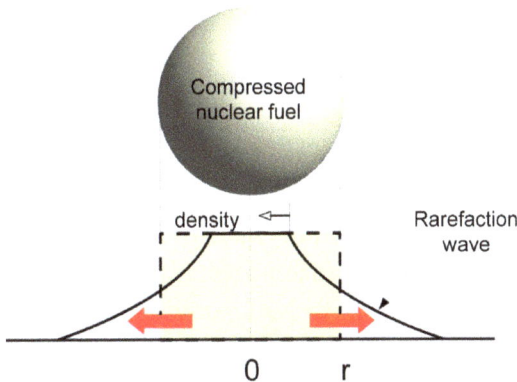

Figure 2.10. A bare sphere of radius r and of an initially uniform density (broken lines) blows off due to a rarefaction wave whose density decreases as the centrifugal velocity increases, the more so the farther from the centre. Its head propagates towards the centre with sonic velocity. Due to the n^2 dependence, most of the thermonuclear burn can be attributed to the dense central zone during the time necessary for the rarefaction to carry away half of the initial mass.

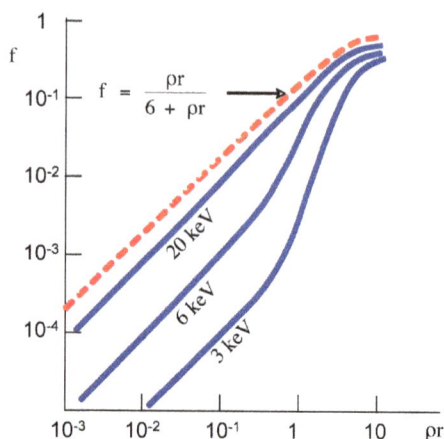

Figure 2.11. Tritium burn fraction f as functions of the areal density ρr for reaction temperatures relevant to an explosive regime. The best conditions correspond to the broken curve. f is then a simple analytic function of ρr.

fusion reaction should surpass the ratio of the plasma thermal energy to the efficiency of the process which produces the high-temperature compressed medium.

Consequently, an efficient inertial fusion system requires an areal density exceeding $2\,\text{gcm}^{-2}$ (at least 30% of the tritium has reacted). Such a criterion is more severe than the Lawson criterion stated for magnetic confinement. Indeed, in inertial confinement, the reaction time is the result of the explosion process, given the mass of the fuel and its density. Since the sound speed depends upon a temperature value that is mandatory (a few tens of keV), the reaction time τ is proportional to r, and ρr is proportional to $n\tau$. Putting numbers in the formulas, it turns out that the required inertial $n\tau$ is about 10 times larger than the magnetic one. This is due to a higher burn efficiency.

For safety reasons, the energy E_F, explosively released by the fusion reaction, should be limited in such a way that the device is not destroyed. A few megajoules is the upper limit (1 megajoule is the energy released by 250 g of TNT, recall). Now, E_F is proportional to the product of f and the mass of the deuterium–tritium mixture. Given the burn efficiency and the related ρr, the mass should be very small. Given the areal density ρr, a similarity relationship is readily obtained for a sphere: its mass is inversely proportional to the square of the mass density (Box 2.4).

Again, putting numbers in the formulas, it is found that $10\,\mu g$ of D–T should have a density 10000 times higher than solid state. As it will be shown in Chapter 7, the compression of matter to such high densities needs an exceedingly high power for a short time, so the corresponding energy remains rather low. Giant pulse lasers with neodymium glass, KrF or CO_2 as active materials, exhibit such properties. Hence, they are used as a primary energy source in inertial fusion.

Box 2.4.

The mass of a sphere of radius r is $M = \frac{4\pi}{3}\rho r^3 = \frac{4\pi}{3}\frac{(\rho r)^3}{\rho^2}$. Given a tritium burn fraction f with given ρr, the similarity relation $M\rho^2 = $ constant is conspicuous. Assume M is $10\,\mu g$ and the areal density ρr is $2\,\mathrm{gcm^{-2}}$. Then, the mass density ρ of the compressed D–T is $1.8 \times 10^6\,\mathrm{kg\,m^{-3}}$, which is close to 10 thousand times the solid-state density (about $210\,\mathrm{kgm^{-3}}$).

References

1. J. L. Tuck, *Nuclear Fusion* **1** (1961) 201; NB. Fusion cross sections are regularly updated.
2. L. Stewart and G. M. Hale, LA-5828-MS (1975).
3. G. Gamow and E. Teller, *Phys. Rev.* **53** (1938) 608.
4. R. F. Post, R. E. Ellis, F. C. Ford and M. N. Rosenbluth, *Phys. Rev. Lett.* **4** (1960) 166.
5. J. H. Nuckolls, L. Wood, A. Thiessen and G. Zimmerman, *Nature* **239** (1972) 139–142.
6. J. D. Lawson, *Proceedings of the Physical Society B* **70** (1957) 6.
7. G. S. Fraley, E. J. Linnebur, R. J. Mason and R. L. Morse, *The Physics of Fluids* **17** (1974) 474.

3

Plasmas

In order to achieve thermonuclear fusion, the active medium should react at a temperature of about 100 million K (10 keV). Well below such temperatures, hydrogen and its isotopes, whose atoms have a single electron, are already completely ionized gases. An ionized material highly sensitive to electromagnetic forces is called *plasma*. Knowledge about this kind of (the fourth) state of matter has improved tremendously over the last 60 years thanks to the research on fusion. Many texts deal with plasma physics. References [1] to [4] are choices among classic or recent books.

1. The plasma state

Scarce on Earth, the plasma state is by far the most abundant state (99%) within the observable universe: it constitutes the whole mass of stars, from the reacting cores to stellar atmospheres and coronas, outer layers of planetary atmospheres, and a large part of interstellar and intergalactic gases. Metals or semiconductors have plasma properties. Our technological societies use plasmas in applications such as electrical discharges, lasers, material processing, deposition of amorphous layers, displays, etc.

Table 3.1. The orders of magnitude of particle densities (n) and temperatures (T) in plasmas. (Figures dealing with magnetic fusion and with the core of an inertially driven target are expected values.)

	Natural plasmas		Laboratory plasmas	
	$n(m^{-3})$	$T(K)$	$n(m^{-3})$	$T(K)$
Interstellar space	10^6	10^4		
Ionosphère (F layer)	10^{12}	10^3		
Glow discharge			10^{17}	10^4
Arc discharge d'			10^{22}	10^5
Magnétic fusion			10^{20}	10^8
Electrons in metals	10^{29}	3×10^2		
Inertial fusion (core)			10^{31}	10^8
Centre of the Sun	10^{32}	10^7		
White dwarfs	10^{36}	10^7		

Due to the long range of forces between charged particles, plasmas exhibit peculiar properties which basically stem from disordered electromagnetism. In Table 3.1, significant examples of natural and manmade plasmas are presented. Prevailing conditions are extremely different from one another. In dense cold media such as the electron gas in metals or the interior of white dwarfs, quantum degeneracy of electrons plays an important role. It also takes part in a transient way in the core of a compressed target aimed at inertial fusion (see Chapter 6). However, most plasmas including fusion plasmas are warmer and less dense. Classical physics are then adequate. So, the question is: which of these properties do all these very different media have in common? A unifying principle lies in the existence of quantities specific to the plasma state.

All plasmas in Table 3.1 contain equal numbers of positive and negative charges. Electrical neutrality is a global property. Locally, excesses of charges of either sign do exist and they generate oscillations and turbulence. The *electron plasma frequency*, often called plasma frequency, is an important quantity (see Box 3.1). Its reciprocal is a time which serves as a reference for all processes involving electrons.

Box 3.1. Electron plasma frequency

The electrostatic force is strong enough to trigger a vigorous reaction to any deviation from neutrality. Electrons carry a negative charge. They are much lighter than positive ions and thus are more easily displaced. They are to oscillate under the return force associated with an excess of positive charges. The angular frequency of these oscillations is given by the formula $\omega_p^2 = \frac{n_e e^2}{\varepsilon_0 m_e}$ where ε_0 is a constant dubbed vacuum dielectric constant. Its presence in the formula is due to the definition of SI units. ω_p is called the plasma frequency. It does not depend upon temperature. This is the eigenfrequency for an electron gas with particle density n_e. It is often associated with resonance effects.

The distance travelled by an electron moving with the mean thermal velocity during time ω_p^{-1} is called the *Debye length* (or radius). As an order of magnitude, this length is the range of the effective electrostatic forces exerted between charged microscopic objects when electron screening effects are accounted for (Box 3.2). This length makes sense if and only if it is larger than the average distance between the centres of force. Most fusion plasmas fulfill this condition. The Debye radius, together with the plasma frequency, is among the natural references for differentiating plasmas.

Box 3.2. Debye radius

In a plasma, the electrostatic potential around a point charge Ze is screened by the electron gas. It exhibits the Debye–Hückel dependence: $\phi(r) = \frac{Ze}{4\pi\varepsilon_0 r}e^{-r/\lambda_{De}}$ where λ_{De} (Debye radius) is the effective range of the potential. For electrons with temperature T, $\lambda_{De} = \frac{v}{\omega_p} = \sqrt{\frac{\varepsilon_0 T}{n_e e^2}}$. This relationship holds only if the electrostatic energy at a distance λ_{De} from the centre of force is smaller than the thermal energy. In order to enforce this criterion, an electron correlation parameter is defined as the ratio of the electrostatic energy at the mean distance between electrons $\langle r \rangle$ to the thermal energy, viz. $\Gamma = \frac{e^2}{4\pi\varepsilon_0 \langle r \rangle T}$. A plasma in which Γ is smaller than 1 is called a dilute plasma. Most fusion plasmas are dilute plasmas.

Other time and length scales are defined for collision processes. Each process has a cross section. The reciprocal of the product of the cross section and the particle density has the dimension of a length. Actually, it represents the average distance travelled by a microscopic object between two collisions, hence the name of *mean free path*. There are as many mean free paths as there are kinds of collisions, including reactive ones. Due to the long-range forces involved, Coulomb collisions need special scrutiny.

2. Coulomb collisions in a plasma

Electron–ion collisions are of paramount importance in a highly ionized medium, but finding a characteristic time is somewhat tricky.

The random walk of an electron within a dilute plasma is quite different from the random walk of a molecule within a neutral gas. Indeed, in a neutral gas, intermolecular forces have very short ranges. Hence, a molecule travels quite a long distance along a line before it is suddenly deviated by another molecule. In a dilute plasma, the Debye radius is much larger than the mean distance between ions. Then, an electron is always under the influence of a scattering centre. Its winding trajectory exhibits only smooth changes in its direction (Figure 3.1).

Thus, in a plasma, microscopic scattering events are ill-defined. As a best educated guess, the electron–ion collision frequency ν_{ei} is associated with a cumulated deflexion of about 90° after multiple Coulomb processes. In fusion plasmas, such a change in direction requires a time much larger than ω_p^{-1}.

Within a plasma, as in other ionized materials, charges (electrons) travel freely. Consequently, the electrical and thermal conductivities are high. These transport processes depend upon collisions between charged particles, mainly electron–ions collisions.

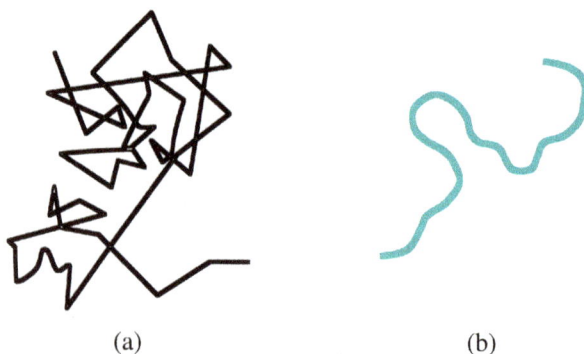

(a)   (b)

Figure 3.1. Comparison of (a) a molecular trajectory within a neutral gas (short-range forces) to (b) the trajectory of an electron perpetually interacting with ions (long-range forces).

As shown by the formula in Box 3.3, the electrical conductivity of a completely ionized plasma is an increasing function of its temperature. Consequently, high-temperature fusion plasmas are comparable with superconductors as far as the resistivity is involved.

The heat transport coefficient κ increases even more rapidly with T. This has a remarkable consequence in the case of thermal waves. The temperature profile does not extend to infinity as it happens with the usual transport coefficients independent of T. On the contrary, the profile ends with a steep frontier separating the wave from the unperturbed medium (Figure 3.2). In

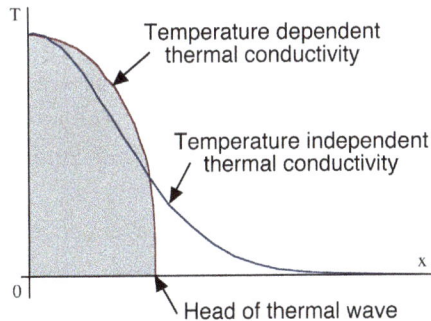

Figure 3.2. Profiles of heat waves. When the transport coefficient does not depend on T, the profile extends smoothly to infinity. When the coefficient is strongly dependent on T, a thermal barrier shows up. At the head of the wave, the temperature rises abruptly.

1D plane geometry, the distance from the source of this *thermal barrier* grows as the square root of the time.

3. Plasma creation

The most direct way to create plasma is heating. As temperature increases, matter is transformed from solid state to liquid, then, to gas, and finally, to plasma. Actually, many other ways exist to reach ionization. For instance, among natural plasmas listed in Table 3.1, metals are in an ionized state at room temperature. Due to the high density, outer atomic orbitals overlap, thus forming a conduction band in which electrons move freely. They form a gas superimposed to the crystalline order of ions. In stellar cores, the hot dense plasma is the result of the adiabatic compression of a gaseous mass under its own gravitation. The simultaneous increase of temperature and density, then, triggers fusion reactions. Now, the extremely rarefied interstellar and intergalactic matter is also ionized: actually, most of it comes from the plasma ejected after violent stellar events. Furthermore, the density is so low that the ionization probability surpasses the recombination probability by several orders of magnitude, even at low temperature.

Lightning is another natural plasma. It involves a transient sinuous channel conducting an intense electric current pulse between two clouds or between a cloud and the ground. The electric field associated with a large potential difference accelerates electrons which are always present in small quantities. Through collisions with air molecules, electrons multiply. An avalanche is created which turns the gas into a conductor.

In the laboratory, electrons from filaments or cathodes are used to ionize rarefied gases. Physicists have studied for a long time gas discharge plasmas. These are created either by applying a high voltage between two electrodes or after a breakdown due to the absorption of electromagnetic waves, e.g. microwaves (Figure 3.3). Other methods use laser light either directly via photoelectric effect, or indirectly whenever a high intensity of radiation interacts with matter. It will be seen in the coming pages which plasma is best suited to each thermonuclear fusion process.

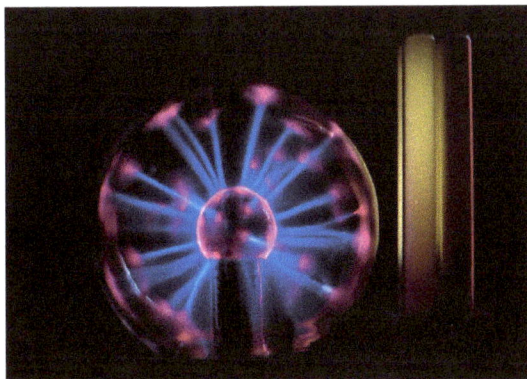

Figure 3.3. Two methods for plasma creation. In a "plasma lamp", plasma filaments result from a high-frequency, high voltage applied between the central sphere and a transparent electrode deposited on the inner face of the glass balloon. Such a device is also an emitter of electromagnetic waves which ionize the gas of an unplugged fluorescent lamp nearby (© J. L. Bobin).

4. The magnetized plasma

Electric currents flowing through plasmas generate magnetic fields. Magnetic fields can also be applied to a plasma with the use of external electric coils.

Whatever its origin, a magnetic field has a strong influence on plasma properties. Indeed, it changes the trajectories of charged particles. The plasma becomes anisotropic. Particle motions are different depending on the direction of the velocity. In the direction parallel to the magnetic filed, the motion is unaffected. On the contrary, in the perpendicular direction, particle trajectories are circular (Figure 3.4). Current loops are thus created with dipole moment opposite to the field: *plasma diamagnetism*. Provided there is no other force, the centre of the circle or *guiding centre* can only move along a line of force.

The angular velocity (Box 3.4) is proportional to the reciprocal of the time needed for a roundtrip. Obviously, in case of high collision frequencies, the particles are deviated before they travel a complete loop. Plasma dynamics are then collision dominated. Otherwise, trajectories are dominated by the magnetic field. The plasma is *magnetized*. The

Figure 3.4. Electrons in a beam injected transversally to a magnetic field move along a circle. The figure refers to a proof of principle experiment. The field direction is perpendicular to the plane of the figure. The beam delivered by an electron gun is made visible thanks to the fluorescence of a rarefied gas it passes through.

magnetization criterion is to be applied to every charged species separately. Electrons in a plasma can be magnetized whilst ions are not, or the other way around. However, in plasmas considered for fusion, all species are magnetized.

Charged particles spiralling around magnetic lines is a simplified picture which holds only if the field is uniform and no other force is present. Should a supplementary force with a component perpendicular to the magnetic field act on the particles, the latter acquire a *drift velocity* orthogonal to both the field and the other force (Box 3.4). The drift velocity is also the perpendicular component of the guiding centre velocity. Many forces can contribute to drifts: electrostatic (Figure 3.5), magnetic-field non-uniformities, etc.

Box 3.4. Charged particles in a magnetic field: relevant parameters

The angular velocity Ω_c of charged particles with mass m and charge Ze spiralling around magnetic lines reads $\Omega_c = \frac{ZeB}{m}$ where B is the magnitude of the field. Ω_c is also called the *gyromagnetic frequency* or *cyclotron frequency*. The *gyroradius* is obtained by dividing the transverse velocity by Ω_c, viz. $R = \frac{v_\perp}{\Omega_c} = \frac{mv}{ZeB}$. At a given velocity, the larger the mass, the larger the gyroradius.

If another force **F** is added to the magnetic force, the charged particle drifts with velocity $v_D = \frac{F \wedge B}{Ze|B|^2}$.

The drift induced by an electrostatic field does not depend upon the sign of the moving charge. Ions and electrons have the same drift velocity. On the contrary, drifts associated with non-uniform magnetic fields have

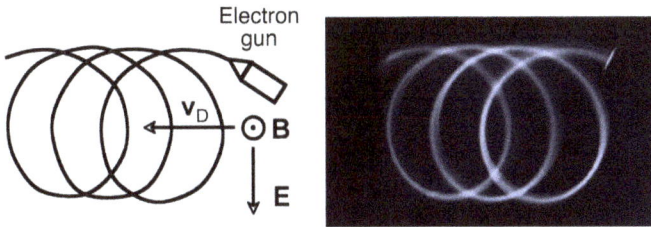

Figure 3.5. Electron trajectories in magnetic and electrostatic fields perpendicular to each other. Electrons undergo a drift in a direction orthogonal to both fields. The drift velocity $\mathbf{v}_D$ is independent of the charge and the mass of the particles; it is the same for electrons and ions in a plasma (after a UCLA document).

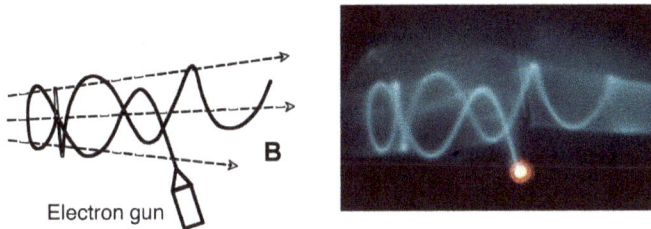

Figure 3.6. Electron trajectories in a magnetic mirror. The magnitude of the magnetic field decreases from left to right whilst field lines are spreading. The guiding centre velocity of the beam emitted by the electron gun decreases and changes sign, so that the beam bounces back towards regions of lower magnetic field (after a UCLA document).

opposite directions for charges with opposite signs. Then, charge separation occurs, hence an electrostatic field and new drifts are created and the process continues.

A special behaviour stems from magnetic field lines getting closer as the field magnitude increases. A charged particle of either sign spiralling around a field line experiences a force parallel to the field line and directed towards the lower field. Consequently, incoming particles are sent back to lower-field regions, this is known as the *magnetic mirror effect* (Figure 3.6). Henri Poincaré [5] performed the first calculation of such trajectories. The effect exists only if the ratio of the perpendicular component to the particle velocity is larger than the square root of the ratio between the minimum and maximum fields.

5. Charged particle diffusion across a magnetic field

In a *highly conducting* plasma, collisions are scarce. The plasma is strongly magnetized. But for the drifts, microscopic charged objects are linked to the field lines. Ionized matter and field lines move together and form the *frozen magnetic field*. Consequently, if part of the plasma is displaced, magnetic field lines follow and get distorted; furthermore, a highly conductive plasma with no magnetic field inside and a magnetic-field-lines system cannot penetrate each other. This is analogous to the Meissner effect, in which a piece of matter becoming a superconductor expels any magnetic field from its interior.

Things are completely different for a *resistive* plasma due to collisions which knock charged particles out of spiralling trajectories around magnetic field lines. These are no longer frozen. Charged particles diffuse through magnetic field lines. Thus, a confined plasma might lose matter and energy. This effect should be kept as low as possible although it cannot be completely controlled.

Transport of particles and energy through magnetic field lines is a problem of paramount importance still under investigation.

Box 3.5. Transport in a magnetized plasma

A simple approach of particle moving perpendicularly to field lines allows one to calculate a *classical* diffusion coefficient which reads $D = \frac{\nu T}{m(\Omega_c^2 + \nu^2)} \approx \frac{\nu}{B^2} \frac{mT}{e^2}$. In the approximate formula for a strongly magnetized plasma ($\Omega_c \gg \nu$), D turns out to be proportional to the collision frequency and inversely proportional to the square of the magnitude of field.

The $1/B^2$ dependence of the diffusion coefficient is favourable to the magnetic confinement. Indeed, calculations show that diffusion would be very slow for a 10^8 K plasma confined by a magnetic field of a few teslas. Unfortunately, in actual experiments, measured coefficients are much higher. Various effects contribute to this result. In many cases, the $1/B^2$ dependence still holds. However, a fairly large multiplying factor should be applied (see Chapter 5).

Plasma waves and turbulence contribute to charged particle leakage through magnetic field lines. The corresponding diffusion coefficient is inversely proportional to the field magnitude (Bohm diffusion). Due to the 1/B dependence, losses are superior than the predictions of classical diffusion, even corrected. Hence, one has to get rid of Bohm diffusion in fusion plasma confinement.

6. Plasma oscillations

Oscillations occur in plasmas. An example was already given on electrons and their characteristic frequency ω_p. Inside a plasma, many waves can propagate and interact. The study of such phenomena is a large part of plasma physics.

The zoo of plasma waves is rich even when no magnetic field is present. It comprises, first, high-frequency modes, i.e. in plasma physics, modes with frequencies larger than the electron plasma frequency ω_p. Transverse electromagnetic waves (including visible light) and longitudinal electron plasma waves (also dubbed Langmuir waves) belong to this category. In Langmuir waves, the electron particle density, the charge separation field and the associated electrostatic potential oscillate together. Both electromagnetic and Langmuir waves propagate only if their frequency is superior to ω_p. The plasma frequency is accordingly a *cut-off* frequency.

An incident electromagnetic wave with frequency smaller than ω_p cannot penetrate the plasma. It is reflected. This is the origin of the brightness and of the reflecting power of metals. Besides being a cut-off, ω_p is a resonance frequency. Contributing to the energy transfer between waves and plasmas, this property is used in the laser interaction with plasmas aiming at inertial fusion. Indeed, a concentrated laser beam is a high-power electromagnetic wave.

On the low-frequency side, oscillatory modes do exist in which ion inertia is involved. Since they look like ordinary sound waves in a gas, they are called *ion acoustic* waves. Electron and ion densities oscillate with the same low frequency.

In the presence of a magnetic field, many types of waves are added to the zoo. Oscillatory modes are classified according to the angles made by the direction of wave propagation as well as the directions of the magnetic field and electric field. Separating high-frequency from low-frequency modes is still relevant; the latter implies ions. Since parallel and perpendicular motions are different, the plasma anisotropy induces, for instance, birefringence. Consequently, there are two cut-off frequencies for high-frequency (HF) waves.

In a magnetized plasma, a special resonance effect appears: there exist frequencies for which a mode does not propagate energy. The stagnation enhances energy transfer from the wave to the plasma. This effect occurs for gyromagnetic resonances dealing either with electrons (frequency denoted as Ω_{ce}) or ions (Ω_{ci}), and for two hybrid resonances. The upper hybrid wave involves electrons only, whereas the lower hybrid waves involve both ions and electrons. All these resonances can contribute to plasma heating. In magnetic fusion machines, ion cyclotron resonance and lower hybrid resonances are mainly used. Indeed, the needed frequencies can be provided with the required power by generators available from industry.

Besides heating, waves influence, for good or for evil, the behaviour of a fusion plasma. The dark side first — turbulence (see next section) can drive locally a break up of magnetic field lines. On the bright side, it was discovered circa 1980 that waves can generate high-intensity currents following the initial current pulse in pulsed machines like tokamaks. Such a *current drive* is a consequence of wave–particle interaction.

As in any other gases, in the electron gas in a plasma, a distribution function represents the way the number of particles is a function of their velocity. The velocity might have any value in the wave propagation direction. In an electron plasma wave, most electrons oscillate longitudinally, except those with a velocity close to the phase velocity. A quasi-steady electrostatic field is actually applied to such electrons that can either be accelerated or decelerated. They look like surfers on an ocean wave (Figure 3.7).

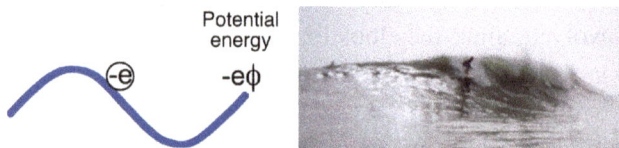

Figure 3.7. Landau effect. Electrons with velocities close to the phase velocity of a longitudinal wave are inside a quasi-steady electrostatic potential. They can be accelerated the same way a surfer on an ocean wave is accelerated by the gravitational potential.

The behaviour of electrons in a plasma wave was investigated correctly for the first time by Landau [6]. Electrons with velocities close to the phase velocity of the wave contribute to energy exchanges between the wave and the plasma particles.

7. Nonlinear waves and turbulence

Plasma properties are complex. Many of them are described by nonlinear equations. This is often the case for oscillations.

As in the case of oceanic surface waves, the propagation of plasma waves is both nonlinear and dispersive. Nonlinearities contribute to the steepening of the perturbation a dispersion tends to smooth out. Whenever both effects compensate each other, a local bump of the electrostatic potential associated with a dip in the density profile might propagate with an amplitude-dependent velocity. Such a solitary wave occurs frequently in space plasmas. It is comparatively scarce in fusion plasmas.

In fusion plasmas, oscillation modes interact with one another. Such couplings might trigger an uncontrolled amplitude growth. This occurs when plasmas are irradiated by electromagnetic waves (radio frequency or laser). Three-modes coupling is the basic mechanism. Given the number of species in the oscillation zoo, many nonlinear effects are expected. Some, such as harmonic generation, are visible to the naked eye. When a powerful infrared laser pulse is concentrated onto a target surface, one can see a green flash emitted by the interaction zone. Besides this spectacular example, other couplings are also noteworthy. When electromagnetic waves are used to heat the plasma, energy might be transferred to the growing waves instead of contributing to the thermal motion of particles, or turbulence could be created. This interesting physics is worth investigating. Furthermore, getting rid of such phenomena is a key issue in some fusion applications.

Other nonlinear effects linked to waves stem from the radiation pressure which can be large every time electrostatic or electromagnetic fields reach high amplitudes. It plays a role in some solitary waves. It also steepens the density profile when a high-intensity laser wave interacts with a plasma presenting a density gradient. Coherent light and a plasma might organize themselves in such away that a diffraction grating that fits the properties of the incident wave is built up.

As in any other gaseous medium, fluctuations take place inside plasmas. Pressure, particle density and temperature exhibit local discrepancies from the average values. Furthermore, in plasmas, electrical neutrality breaks down locally, thus generating random electric fields. Turbulence in plasmas is made of waves that grow, damp, and interact in complete disorder. There is no such thing as a quiet hot plasma. The wavelength spectrum and the way energy is transmitted from mode to mode are important knowledge, so are the contributions of different kinds of turbulence to heat transport. Incident electromagnetic wave energy might go to disordered oscillations rather than to plasma heating. As another example, Bohm diffusion across magnetic field lines is driven by a turbulent regime.

8. Natural and fusion plasmas

The high temperature plasma is unstable, full of nonlinear effects and leaks easily from any attempted confinement. In nature, a huge mass effect keeps this elusive medium inside stellar cores, the place where fusion energy is released. A terrestrial device needs to work differently.

Man-made fusion devices mimic natural processes although not completely. Natural plasmas and plasmas intended to be produced in fusion machines can be compared on a single diagram (Figure 3.8) in which the abscissa is particle density and the ordinate is temperature.

The Universe as we know it is populated with orbiting objects far from one another. They move within a "vacuum" which is actually an extremely rarefied gas. Since ionisation mechanisms are more likely than recombination, the gas is ionized. Plasma extends from the centre of stars to intergalactic spaces.

Due to the enormous gravitational attraction associated with the mass of the Sun, the conditions prevailing inside the core are expected to last for the whole stellar life. In the solar corona, the density is low with a temperature much higher than at the stellar surface. Beyond the corona, both temperature and density decrease as plasma flows away to interstellar space (solar wind). Nebulae are made of dilute dust and mostly ionized hot gases.

Down on Earth, in order to achieve D–T fusion, the relevant temperature is 100 million K (10 keV) higher than in the solar core (15 million K).

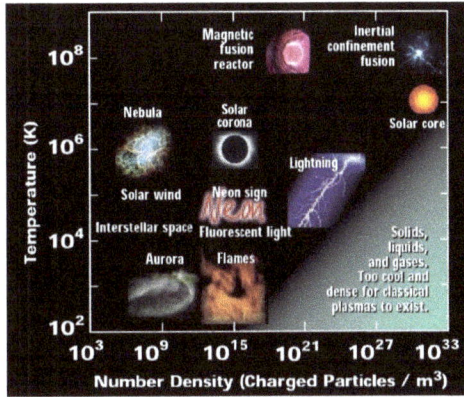

Figure 3.8. (after a CPEP document). Locations of natural and fusion plasmas on a temperature–density diagram (logarithmic plot). Conditions in magnetically confined plasmas (10^8 K, 10^{20} m^{-3}) are far from natural plasmas. In inertial fusion, temperature and density (10^8 K, 10^{32} m^{-3}) are close to solar core conditions (1.5×10^7 K, 10^{33} m^{-3}).

Quasi-steady (magnetic confinement) and explosive (inertial confinement) regimes involve different densities. In magnetic confinement, an upper limit is imposed by the strength of the structural materials. The power density of the fusion reaction should not exceed some $100\,\mathrm{W/cm^3}$. The maximum particle density is, then, 10^{20} per m^3. Such conditions are far from natural plasmas. On the contrary, the required temperature and density for inertial fusion are close to the conditions prevailing in stellar cores.

To achieve magnetic fusion, it is necessary to get a steady state of matter that is transient in nature. Conversely, in inertial fusion, the reaction occurs in a transient state that is stabilized in nature thanks to an intense gravitational field. In both cases, enormous difficulties are to be met.

References

1. L. Spitzer, *Physics of Fully Ionized Gases*, 2nd ed, Wiley (1962).
2. F. F. Chen, *Introduction to Plasma Physics and Controlled Fusion*, 2nd ed, Plenum Press (1984).
3. R. J. Goldston and P. H. Rutherford, *Plasma Physics*, IOP (1995).
4. J. M. Rax and C. Guthman, *Physique des plasmas*, Dunod (2005) in French.
5. H. Poincaré, *C.R.A.S.* **CXXIII** (1896) 530.
6. L. D. Landau, *J. de Phys. URSS* (1946).

4

Some Features of Magnetic Confinement

1. The basics

It was shown in the preceding chapter that in a magnetized plasma, charged particle trajectories spiral along field lines. This effect links matter to the magnetic field. An alternative approach follows from the macroscopic equilibrium of gaseous matter: In the absence of external forces, matter stays motionless provided the pressure is uniform. A similar condition should also apply to the magnetized plasma (Box 4.1).

Box 4.1. Energy densities and pressure

Let p be the pressure of the ionized gas in a magnetized plasma. The magnetic pressure is equal to the field energy density, i.e. $\frac{B^2}{2\mu_0}$ where μ_0 is a constant imposed by SI units. Then, an equilibrium implies the spatial uniformity of the total pressure, hence the condition: $p + \frac{B^2}{2\mu_0} =$ constant. Consequently, a low-density, high-temperature magnetized plasma ($p \neq 0$, 10^{20} particles per m^3, 10^8 K) can be isolated ($p = 0$) from any external wall by a magnetic field whose pressure is the sum of plasma and magnetic field pressures in the inner region.

Since 1950, many magnetic configurations have been investigated. The simplest confinement device is the *mirror machine*. Experiments have been performed with this device since the early beginnings of controlled fusion research. Basically, a magnetic trap is made of two coaxial magnetic mirrors. The plasma is expected to stay in the central region of lower magnetic field. However, there is a major drawback. Obviously, a particle with a velocity along the field lines is insensitive to the magnetic force and leaks through the mirror. More precisely, particles travelling towards a mirror are reflected back provided the component of their velocity parallel to the field is not too large with respect to the perpendicular component. Thus, depending on the direction of their velocity with respect to the field lines, the particle trajectories are either passing or trapped. In order to reduce the axial losses in mirror machines, plug barriers were designed and implemented. Such devices are quite complicated and did not prove very efficient. Mirror machines are no longer considered as promising.

2. Current-carrying plasmas

Since a plasma is a good electric conductor, it can contribute to its own confinement whenever it carries a current. The subsequent magnetic field applies on the local current a Laplace force orthogonal to both the field and the current provided field and current lines are not parallel. An equilibrium might hold when this force opposes a pressure gradient in the ionized gas.

Box 4.2. Magnetic (or flux) surfaces

In a motionless, confined plasma, the magnetic Laplace force $\mathbf{j} \wedge \mathbf{B}$, where $\mathbf{j}$ is the local current density, compensates exactly the local pressure gradient ∇p, so that $\nabla p = \mathbf{j} \wedge \mathbf{B}$. Consequently, such an equilibrium implies $\mathbf{j} \cdot \nabla p = 0$ and $\mathbf{B} \cdot \nabla p = 0$. Hence, current lines together with field lines lie on constant-pressure surfaces orthogonal to the pressure gradient. Since they enclose constant-flux tubes, such surfaces are called *magnetic flux surfaces* or, more simply, *flux surfaces*. Their shape depends on the geometry of the confinement device — plasma inside magnetic field lines.

A cylindrical column carrying an intense axial current is the simplest case of a self-confined plasma. In this axisymetric situation, magnetic surfaces are coaxial cylinders. The magnetic field is azimuthal and increases from zero at the axis to a maximum value at a radius close to the radius of the plasma surface. It decreases out of the plasma. The Laplace force inside

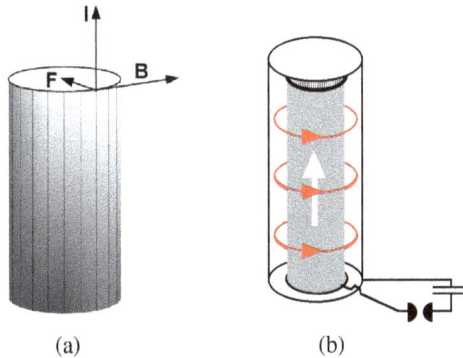

Figure 4.1. (a) The Laplace force **F** from the magnetic field **B** on the current **I** is directed towards the axis. (b) Scheme of the pinching of a plasma column (Z pinch). The discharge of a capacitor bank produces the high-intensity current. A return current flows along a metallic cylinder surrounding the plasma column.

the plasma points inwards, whereas a pressure gradient acts in the opposite direction. A steady state might be obtained. Then, the plasma temperature, provided it is uniform, is proportional to the square of the current intensity (Bennett).

Now, a dynamical process has to be used in order to build up a hot plasma column. A few tens of kilovolt are applied between two coaxial circular electrodes to a low pressure gas. The gas is spontaneously ionized, and a few tens of kiloampere from a capacitor bank can flow through the just-created plasma (Figure 4.1). At the onset of the process, the current greatly exceeds the equilibrium value. The column contracts under the magnetic force. Such a *pinch* effect ("Z pinch" or linear pinch) is a generalisation of the attraction, known since the days of Ampere, between two parallel currents.

Pinching of the plasma column induces a strong heating. The temperature obtained that way might be of interest for fusion. However, since it results from a violent process, the hot plasma is in a transient state. The maximum compression is followed by a rapid expansion. Furthermore, instabilities arise which result in blowing off the plasma column.

Worse, the electrodes are a solid surface in contact with the plasma, thus driving serious heat losses. Consequently, a Z pinch cannot be foreseen as an efficient device for magnetically confined fusion. However, it might be used as a powerful radiation source suited for inertial fusion (see Chapter 8).

3. Stability

The state of a magnetically confined plasma depends upon a host of variables which belong to several domains: thermodynamics, electromagnetism, mechanics, geometry, etc. A stable equilibrium should correspond to a function of all these variables, namely a potential energy. Simply stated, the principle leads to cumbersome calculation. Indeed, the number of variable is large; the number of potential instabilities is even larger. There is no such thing as a general stability criterion. Investigating stability obeys Murphy's law[2] since

- The system is stable provided the potential energy increases for every possible perturbation.
- If the potential energy decreases for a single perturbation, the system is unstable.

In practice, every kind of perturbation is investigated separately by checking the stability of the equilibrium and finding ways of mitigating the instability. The stability problem is of paramount importance for scientists involved in controlled fusion research. It arises even in the simplest cases.

Consider a current-carrying plasma column. It can readily be shown that its equilibrium is not stable. Indeed, the plasma boundary is the outmost magnetic surface. Now, from the laws of electromagnetism, the magnitude of the magnetic field at the boundary varies as the reciprocal of the square of the radius. Imagine an axisymetric distortion such as a local narrowing (Figure 4.2 a). At this point on the plasma surface, the magnetic pressure is much higher than the plasma pressure, which is uniform on the whole surface. The imbalance drives further narrowing. By the same token, in a local swelling, the magnetic pressure is lower than the plasma pressure and the defect is to grow.

Such a distortion in which narrowing alternates with swelling is dubbed "sausage" or in scientific language, a zeroth order instability. In a cylindrical column, instabilities are labelled by a number m accounting for the complexity of the initial defect. Order 1 is a "kink", in which the column

---

[2]"If anything can possibly go wrong, it will!"

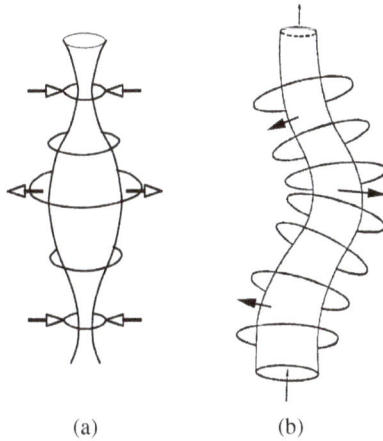

(a)   (b)

Figure 4.2. Instabilities of a plasma column. Distortions of a magnetic surface: (a) "sausage"; (b) "kink".

is locally off centre. In regions where field lines are closer to each other, the magnetic pressure is higher than in regions where they are far apart. Again, the perturbation increases (Figure 4.2 b).

Higher orders refer to more complex braided geometries.

The above examples are *MHD*[3] *instabilities*. It has been known since the earlier experiments that they can grow until the plasma column is blown off in a time much shorter than the minimum confinement time required by the Lawson criterion (see Chapter 2).

A few recipes can be used to overcome such instabilities. Their principle is rather simple. Consider, for instance, the kink instability of a plasma column inside a coaxial conducting cylinder. When the plasma is distorted, the magnetic field lines are displaced. The magnetic flux across meridian planes varies and inside the conducting wall, this creates eddy currents that oppose the flux variations. The column is thus stabilized.

Another stabilizing method uses a longitudinal magnetic field created by an external coaxial solenoid. Field lines are helices combining azimuthal and longitudinal fields. Now consider a helical spring. The more it is stretched, the more difficult it is to move it sideways. By the same token, a stability criterion for a magnetically confined plasma column requires a

---

[3] for Magnetohydrodynamics

longitudinal component much larger than the axial one. Consequently, the ratio of the plasma pressure to the magnetic pressure, β is smaller than 1, typically a few percent.

4. Closed configurations

Contact with electrodes induces a vigorous cooling of the plasma in a linear discharge. This is a major drawback that can be avoided by an electrode-less plasma ring confined by appropriate magnetic field lines. The archetypal geometry of a closed magnetic configuration is the torus with a circular meridian cross section (Figure 4.3). The ratio of the major radius R_0 to the minor radius a at the plasma boundary defines the *aspect ratio*. This ratio still makes sense for non-circular meridian cross sections, as it is often the case in actual devices.

When a current-conducting wire is winded onto a torus, the magnetic field lines are circles whose centre lies on the torus axis. This toroidal field is denoted by B_T. Contrary to the case of a linear solenoid, its magnitude is variable. It is inversely proportional to the distance from the torus axis. A plasma ring can be accommodated in this magnetic configuration since diamagnetism reduces the overall field magnitude inside the plasma.

Due to the curvature of the toroidal magnetic field lines and to the spatially varying magnitude, drifts contribute to plasma leakage through

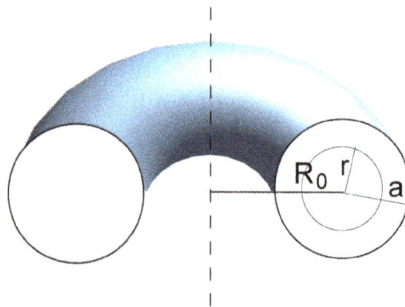

Figure 4.3. The toroidal plasma ring with circular meridian cross section. The centre of a meridian cross sectional circle lies on a circle of radius R_0, the major radius of the torus. a is the minor radius at the plasma boundary. R_0/a is the aspect ratio.

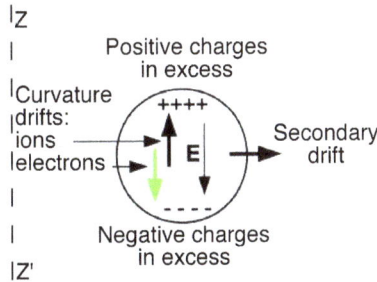

Figure 4.4. Drifts in a magnetically confined plasma torus (toroidal field only). The non-uniformity of the magnetic field induces drifts parallel to the torus axis Z'Z. Electrons and ions drift in opposite directions. Positive charges accumulate in the upper parts of the ring, whist negative charges accumulate in the lower parts, thus, creating an electrostatic field and a subsequent secondary drift, which drives the particles outwards.

a two-step mechanism (Figure 4.4). Eventually, particles are driven outwards.

Winding magnetic field lines onto toroidal flux surfaces precludes the occurrence of such drifts. Then, the magnetic field still has a toroidal component to which a *poloidal* component B_p in meridian planes is added. Now, consider intersections of a magnetic field line with a meridian plane. Successive intersection points lying on a circle are separated by an angle ι, which defines a *rotational transform*. As Andrei Sakharov had shown in the early 1950s [1], a rotational transform can be realized if the plasma ring carries a toroidal electric current with no dynamical contraction. In the late 1960s, a device was operated in Moscow according to this principle with a comparatively small aspect ratio of 3 [2]. Dubbed *tokamak*, it proved rid of the Bohm's diffusion (see Chapter 3), which had driven unacceptable plasma losses in most devices implemented since 1950. The tokamak with an inductively generated current is archetypal of a magnetically confined plasma torus.

In an alternative approach, the rotational transform is obtained without currents by applying external magnetic field with a suited geometry. This is done in plasma rings with complicated shapes called *stellarators*. Such configurations are still in use. However, presently operating devices (see below, Section 9) are smaller and less advanced than third generation tokamaks.

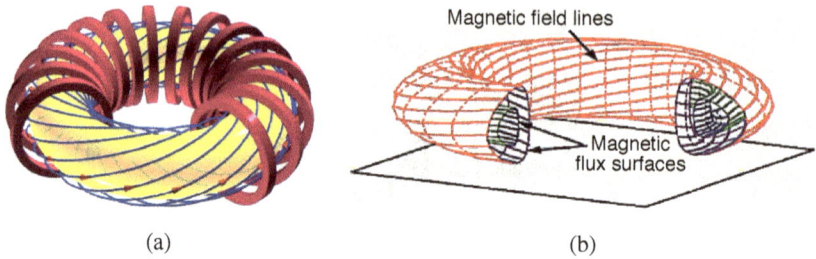

Figure 4.5. (a) Current-carrying plasma ring inside a toroidal solenoid. Field lines are winded onto toroidal surfaces; (b) Magnetic flux surfaces are tori nested one inside the other.

5. Equilibrium and stability of a current-carrying plasma ring

In a tokamak, i.e. a current-carrying plasma ring inside a toroidal solenoid, the magnetic field has two components: the poloidal component B_p due to the current itself and the toroidal component B_T due to external coils. Both contribute to the confinement: B_p reduces the leaks associated with the toroidal geometry. Furthermore, locally, the corresponding Laplace force compensates the pressure gradient. Now, B_T contributes to the MHD stability the same way as in the linear pinch provided it is much larger than B_p. Consequently, the ratio β of the plasma pressure to the total magnetic energy density is at most 10%. In real life, it stays below 5%.

Another consequence deals with the geometry of field lines: they spiral on magnetic flux surfaces, which are tori nested one inside the other. The innermost flux surface reduces to a circle called the *magnetic axis*. Since the pitch of the winding is different from one flux surface to another, the field is said to have *shear*. The MHD stability is improved with increasing stiffness of the field, i.e. the number q of roundtrips a field line makes around the torus axis while completing a single poloidal turn along a minor circle. The number q is called the *safety factor*. It is constant for a given flux surface and is linked to the rotational transform of the angle ι by the formula $q = 2\pi/\iota$. A rational q corresponds to closed field lines with consequences on the onset of some other kinds of instabilities (see below, Section 7). In general, a good stability is achieved when q is larger than 1 in the whole plasma ring. Unfortunately, it is not always possible to control this parameter. Usually,

Chapter 4. Some Features of Magnetic Confinement

its value in the vicinity of the magnetic axis is slightly smaller than 1 and increases up to 3 at the plasma boundary. A method to shape the current profile in order to obtain prescribed values of q will be presented in the following chapter.

The metallic conductors and the structural materials experience huge magnetic forces. Hence, the magnitude of the total magnetic field is to stay below an upper limit inside the range 5–10 tesla, depending on the particular design. An upper limit also applies to the poloidal component at the plasma boundary and to the ratio of the total intensity I to the minor radius a (see Box 4.3). A large current implies a large area. Aspect ratios in tokamaks are, thus, smaller than 3. The typical tokamak looks more compact than other ring-shaped devices.

Box 4.3. Safety factor and limiting conditions

The safety factor is the number q of roundtrips a field line makes around the torus axis while completing a single poloidal turn along a circle with minor radius r. It reads $q = \frac{rB_T(R_0)}{R_0 B_p(r)}$ where R_0 is the major radius of the torus (radius of the magnetic axis).

To a first but crude approximation, the torus is considered as a cylinder whose radius is the minor radius at the plasma boundary, and the azimuthal field replaces the poloidal field. Then, $B_p(a) = \mu_0 \frac{I}{2\pi a}$, and elementary electromagnetic formulas linking the plasma pressure, the azimuthal magnetic field, the radius and the current intensity are extended to the toroid case. The condition on the total magnetic field implies an upper limit for the poloidal component. $\beta \ll 1$ implies, for the pressure and the current intensity, the following inequalities:

$$ p \ll \frac{qR_A}{4\mu_0} \frac{B_T I}{a}, \quad I \ll \frac{\pi\sqrt{2}}{\mu_0^{3/2}} aB_T $$

where $R_A = R_0/a$ is the aspect ratio. In practice, upper limits are defined, and the above inequalities are replaced by equalities with semi-empirical proportionality coefficients. Such formulas prove useful in the design of a tokamak.

Another imbalance effect stems from the toroidal geometry. Indeed, a magnetic surface experiences a force resulting from the plasma kinetic pressure. The force is smaller on the part facing the torus axis (smaller area) than on the part facing the opposite direction (larger area). The imbalance induces a radial swelling which is not compensated by the magnetic field, whose magnitude decreases as the distance from the axis increases.

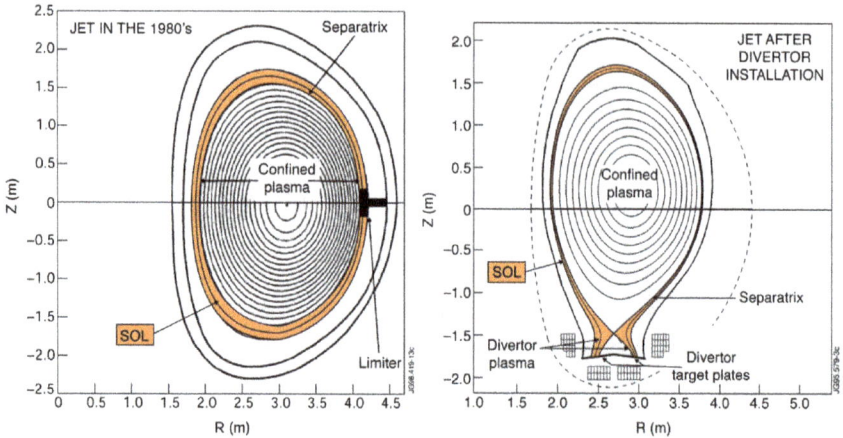

Figure 4.6. Meridian cross sections of a tokamak (JET). Plasma boundary is defined either by (a) a limiter or by (b) a flux surface with an X point (separatrix). In both cases, the confined hot plasma is surrounded by cold plasma with impurities that are driven towards either the limiter or target plates beyond the X point (the principle of the divertor). SOL stands for Scrape Off Layer.

Other forces should act on the plasma in order to stop the swelling. This occurs automatically in small devices in which the plasma ring is created inside metallic walls with a gap to avoid toroidal currents. The mechanism was described in Section 3 for the case of the linear pinch. It proves insufficient for bigger machines. Then, a compensating vertical magnetic field is applied to the conducting plasma ring so that the subsequent Laplace force opposes the pressure imbalance. The magnetic field geometry is modified and includes a special magnetic surface with at least one cross point (X) (Figure 4.6). Obviously, the confinement holds only inside this surface, called a separatrix. The separatrix is the natural plasma boundary, better than the boundary resulting from a limiter that contributes to conductive thermal losses. In any case, there is no such thing as a sharp plasma boundary.

The magnetically confined plasma is not completely insulated from the walls of the vessel inside which it was created. In the external colder layers of the plasma ring, specific instabilities occur: Edge Localized Modes (ELM). They contribute to the acceleration of ions, which eventually strike the wall and induce sputtering. Thus, multiple charged ions penetrate into the plasma thanks to such plasma–wall interactions. Since they emit intense line radiation and enhance bremsstrahlung (proportional to Z^2), these impurities are detrimental. In order to eliminate impurities, the outer plasma layers (Scrape Off Layers, SOL) are in contact either with a limiter or with absorbing plates towards which they are guided via a device called a *divertor*. Every big tokamak device includes a divertor.

6. Charged particle trajectories in a current-carrying plasma ring

In a tokamak, the dominant toroidal component of the magnetic field has a magnitude inversely proportional to the distance from the torus axis. A mirror effect is expected whenever a charged particle comes closer to the axis.

Passing particles have velocities which do not fulfill the mirror condition. They spiral indefinitely around the magnetic axis. Consider a meridian plane co-rotating with the guiding centre around the torus axis. In this plane, the trajectory of the guiding centre is a circle whose centre is slightly offset from the magnetic axis. This motion contributes to the macroscopic current which has a poloidal component. Current lines spiral on the magnetic surface. In the vicinity of the magnetic axis, pressure gradient are expected to be small, so has to be the opposing Laplace force, implying current lines have a direction close to the direction of magnetic field lines. A vanishing angle would mean no Laplace force at all and a uniform pressure.

Trapped particles travel back and forth between two mirrors symmetrically located with respect to the equatorial plane. In the co-rotating meridian plane, the guiding centre trajectory looks like a banana, hence the name given to this particular behaviour. Trapped and passing trajectories are represented in Figure 4.7.

For the sake of completeness, some particles (a negligible minority) have guiding centre trajectories parallel to the axis of the torus. They do not participate in the confinement.

The trapping of charged particles has an interesting consequence. Consider two neighbouring bananas (Figure 4.8). The innermost B_1 is located in a higher-pressure, higher-density region. Hence, more particles

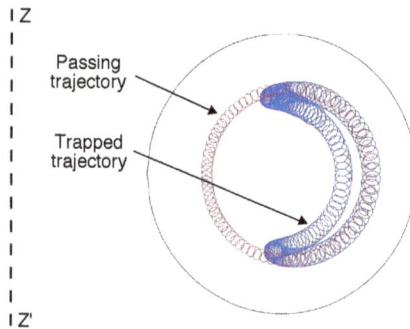

Figure 4.7. Charged particle trajectories in a tokamak as seen in a meridian plane co-rotating with the guiding centre around the axis Z'Z (numerical simulation). A passing trajectory is coloured in red; a trapped trajectory called the "banana" is coloured in blue.

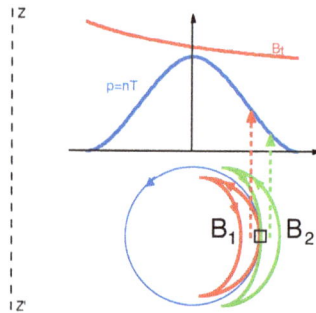

Figure 4.8. The generation of the bootstrap current. Due to the density profile, more particles are trapped in banana B_1 (red) than in the adjacent banana B_2 (green). In the region where the bananas are tangential, the electric current density is thus enhanced.

are trapped by collisions in this banana than in adjacent banana B_2 located in a lower-density part of the plasma ring. In the volume where B_1 and B_2 are tangential, trapping enhances the current flowing along the ring. This self-generated current is dubbed "*bootstrap*", a reminiscence of Baron Munchhausen lifting himself up by his bootstraps. The bootstrap current contributes to the shear and helps to prevent turbulence.

Collisions also expel particles from trapped trajectories, the more so for particles counter streaming with respect to the main current. The detrapped particles are to drift according to the mechanism presented in Section 4.4 (Figure 4.4) and contribute to diffusive leakage. The corresponding particle and energy transport is called *neoclassical* since it remains with a $1/B^2$ dependence. Three regimes of neoclassical transport are classified according to the values of the collision frequency (Box 4.4).

7. More instabilities in the torus

Specific instabilities take place in toroidal plasma rings. An example was quoted in Section 4.5: the radial swelling that has to be compensated by an additional magnetic force. By the same token, instabilities are driven by local pressure imbalance whose dynamics look similar to the lifting of a hot hair balloon, hence the name "ballooning modes".

Other instabilities are associated with the values of the safety factor q. Whenever this ratio is rational, it corresponds to closed field lines. A closed field line can accommodate a whole number of wavelengths of a periodical distortion. This property also holds in the orthogonal direction on the same magnetic flux surface. Similar to air column resonances in organ pipes, resonances can take place on flux surfaces with integer or rational q.

Then, for instance, internal modes of instability might grow in the vicinity of flux surfaces with rational q. For such modes, the magnetic shear is a stabilizing factor. The confinement is stable provided q > 1.

Kink modes associated with the distortions of the magnetic axis are another example: the centre might move off-axis; the shape might vary from circular to elliptical; higher order defects might appear. These instabilities have a positive growth rate when q is either smaller than one or has a value slightly below other integers and rationals.

Now, the most spectacular instability associated with a rational q is the *tearing mode* (Figure 4.9). Plasma resistivity and magnetic shear contribute

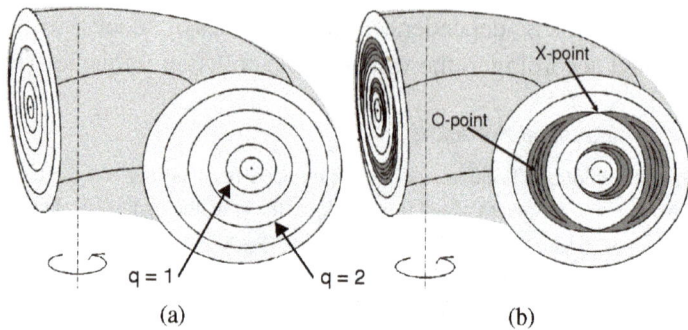

Figure 4.9. Tearing modes in a tokamak. (a) Magnetic shear together with the plasma resistivity induce an instability which takes place on flux surfaces for which q is a whole or rational number. As in the case of an air column, a resonance corresponds to a whole number of wavelengths along a closed field line. (b) The associated flux surfaces are split by the growth of a radial component of the magnetic field. Magnetic islands for $q = 1$ and $q = 2$ are shown on a meridian cross section.

to major changes in the magnetic field and current distributions. A local geometry in which one component of the magnetic filed is uniform whilst the orthogonal component changes sign across a neutral line is unstable. The resistivity of the current-carrying plasma drives a third component, orthogonal to both initial components, which grows periodically along the neutral line.

As a consequence, magnetic islands appear carrying currents flowing in a direction opposite to that of the initial current. In meridian cross sections, magnetic islands are enclosed within a separatrix, i.e. a flux surface that exhibits X points. A magnetic island is associated with a plateau in the particle density. In the vicinity of X points, the nonlinearity of field–particle interactions introduces disorder in the magnetic field map, which might evolve to a chaotic state, thus contributing to an anomalous leakage of particles. Furthermore, the growth of magnetic islands can produce local short circuits called *disruptions*, which blow off the geometry of magnetic field lines and plasma. A *major disruption* would involve the whole plasma ring and lead to catastrophic destructions. Although a few cases of such events have been reported, they are very unlikely.

In a magnetically confined plasma ring, tearing modes with slow growth rates might occur for flux surfaces with $q = 1$ (a single magnetic island) or

$q = 2$ (two islands). For tearing modes, the configuration is stable only if $q \geq 3$. However, tearing modes with a very slow growth can be accepted to a certain extent.

8. Compact tori

The higher the q, the better the stability of the magnetized plasma ring. High q (typically greater than 10) are obtained in plasma rings with an aspect ratio close to unity. The ring is then closer to the torus axis, and field lines make many turns on the part of the flux surface facing the axis, which in this case looks like a long hollow cylinder. In addition to stability, there are further advantages such as the device is very compact and the toroidal magnetic field can be created by an axial conductor.

Compact tori are investigated at Culham (England) and Princeton (New Jersey, USA). In both plasma physics laboratories, the development strategy is the same as that for tokamaks: successive generations of increasing size. Presently, experiments are carried out on medium-sized devices with currents of order one MA, e.g. MAST [4] in Britain and N.S.T. in the US.

In ordinary plasma rings such as tokamaks, the ratio β of the plasma pressure to the field energy density, is at most 5%. In compact tori, β is a few tens of percent. If it could reach unity, simple, compact and comparatively cheap thermonuclear reactors could be achievable.

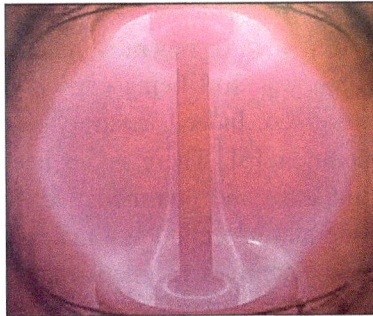

Figure 4.10. The plasma in a compact torus (MAST). The toroidal field is created by an intense current in the axial conductor.

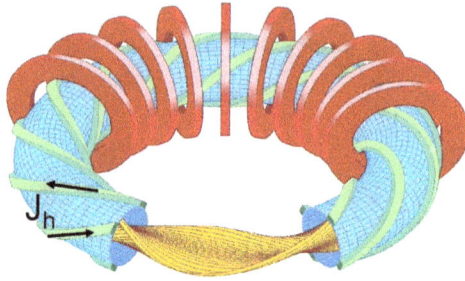

Figure 4.11. (MPG/IPP Greifswald document). A simple stellarator design. Added to the coils creating the toroidal magnetic field are four helical circuits wound onto a torus carry currents J_h with alternatively opposite directions. Meridian cross sections of flux surfaces vary along a major circle from an elongated ellipse to a rounded triangle and back with period $2\pi/5$.

9. Stellarators

Specially designed magnetic coils are an alternative way of implementing a rotational transform. Such devices were invented by astrophysicist Lyman Spitzer [5] in 1951 and named *stellarators*. The Princeton Plasma Physics Laboratory was established in 1952 to operate stellarators and did so for about 15 years. Afterwards, the laboratory converted to tokamaks.

The original stellarator has the figure of a racetrack, with drifts in successive U turns hopefully compensating each other. Later, another simple design was introduced, adding an even number of helically winded coils carrying currents in opposite directions to the coils generating the toroidal field (Figure 4.11).

For many years, the stellarator was rated much less than the tokamak. However, some research was still being carried out and recently, new important programs resumed. Indeed, contrary to tokamaks, stellarators allow continuous operation [6]. There is no need for intense current generation — both inductive and non-inductive. Furthermore, advances in numerical computing and in technology make it possible to precisely design and implement the complex magnetic maps that are necessary for an efficient confinement.

So far, stellarators are not as big as tokamaks nor are operated by large international teams. Research programs are still active in the U.S.A., Japan

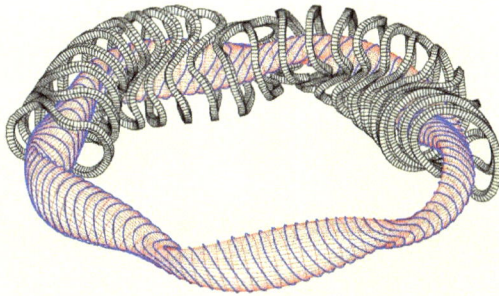

Figure 4.12. (MPG/IPP Greifswald document). Wendelstein 7-X: the geometry of the superconducting coils and of the plasma boundary from numerical computing. Poloidal current lines are drawn on the plasma surface, whose involved shape matches the complexity of the magnetic field map.

and Germany, where "Wendelstein 7-X" is being built at the Max Plank Institute in Greifswald (Figure 4.12). In this device, the magnetic field map is created by superconducting coils with complicated shapes. The plasma ring is pentagonal with a high aspect ratio. The major radius is about 5 m. After cumulative delays due to difficulties in coil machining, the assembly of Wendelstein 7-X will be completed in 2014 [7].

References

1. English translation, I. E. Tamm and A. D. Sakharov, Magnetic thermonuclear reactor theory, in *Plasmas Physics and the Problem of Controlled Thermonuclear Reactions*, M. A. Leontovich, ed., Vol. 1, Pergamon (1961).
2. L. A. Artsimovich, Tokamak devices, *Nuclear Fusion*, **12** (1972) 215.
3. D. Pfirsch and A. Schlüter, Rapport de l'Institut Max Planck MPI/PA/7/62.
4. http://www.ccfe.ac.uk/MAST.aspx.
5. Lyman Spitzer, *A proposed stellarator*, PM-S-1 (1953); *Phys. Fluids* **1** (1958) 253.
6. F. Wagner and H. Wobig, The stellarator reactor, *Landolt–Börnstein Handbook on Energy Technologies*, Band VIII/3 (2000).
7. http://www.ipp.mpg.de/ippcms/eng/pr/forschung/w7x/index.html.

5

Tokamaks

In the late sixties, it was proven that, in tokamak T-3 at the Kurchatov Institute in Moscow, hot plasma could be magnetically confined for durations that other devices were unable to achieve. With this result came an improved confidence about magnetic fusion. The way was open for important research and development programs to be carried out in the USSR (for as long as this entity existed), the USA, Japan and Europe.

1. The description and operation of a tokamak

All tokamaks built so far are experimental facilities. They accommodate a host of tuning devices and instruments aimed at diagnostics and measurements. In the biggest installations, a dedicated power plant is necessary to feed magnetic field coils and current generation. Altogether, tokamaks are quite complex.

The toroidal current is inductively generated. Therefore, the tokamak is a pulsed device. The plasma ring acts as the secondary coil of an electrical transformer. The coupling with the primary coil is often improved by a magnetic circuit made of laminated steel. In big systems, the mass of this component is a few hundreds to a few thousands tonnes (e.g. 800 for Tore Supra; 2500 for JET).

Before starting the current pulse, the toroidal vessel is evacuated to an exceedingly low pressure in order to minimize the impurity content of the gas introduced for a given experiment: hydrogen, deuterium, helium. Indeed, the initial gas density ranges from 10^{19} to 10^{20} m^{-3}, i.e. a pressure below 1 Pa (10^{-2} torr) at room temperature. Whilst toroidal field and poloidal field coils are fed with quasi-continuous current, the primary coil of the transformer is connected to a high-power pulse generator. The induced electric field is high enough to accelerate free electrons (some are always present in any gas) and trigger an avalanche which eventually leads to complete ionization of the gas, transforming it into an electric conductor. The current pulse in the secondary coil has a trapezoidal time history with linear leading and trailing edges. The leading edge should be slow enough to prevent dynamical pinching of the plasma ring. The length of the plateau is adjusted to the expected duration of the confinement.

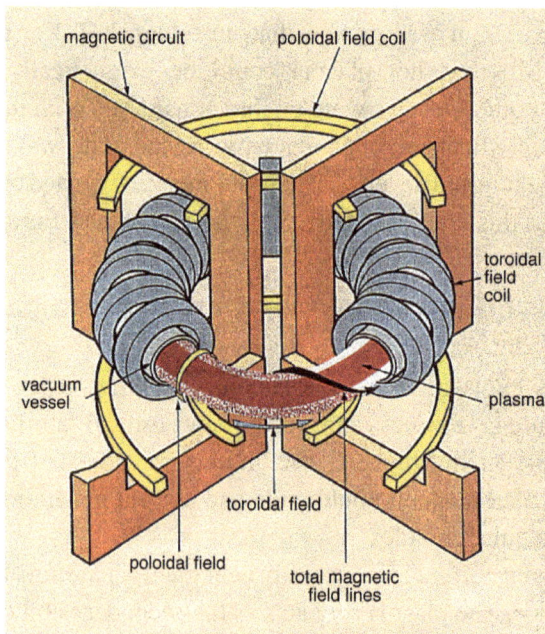

Figure 5.1. A scheme of a tokamak. (From a CEA document) The vacuum chamber which is to accommodate the plasma ring is surrounded by toroidal field coils. The plasma ring is a single-turn secondary circuit or a transformer. Some parts of an optional magnetic circuit are sketched; it might improve the coupling with the primary circuit (not represented). The main current is inductively generated by applying pulsed power to the primary circuit.

Since the pulsed current takes place spontaneously in the plasma, it is almost impossible to prescribe a given value of the safety factor q in the vicinity of the magnetic axis. In experiments, this value is in the range 0.8–1. As a function of the minor radius, q increases quadratically. In order to increase q on the magnetic axis, the current distribution across the plasma ring should be modified. It will be shown later how to achieve a given q profile.

2. Bigger and bigger, why?

The early tokamaks had modest sizes. As new devices were designed and implemented, tokamaks grew bigger and bigger from one generation to the next. Meanwhile, the number of machines per new generation decreased as the costs were rising. The latest generation concentrates on a single international project: ITER. This evolution follows from semi-empirical *scaling laws* which all predict better confinement in larger plasma rings.

In thermonuclear research, achieving high temperatures, millions and even tens of millions degrees, proved quite easy and occurred early. Satisfying the Lawson condition at these temperatures is a much more difficult challenge. In the $n\tau$ product, the particle density n is bounded from above by the maximum steady magnetic field that can be created in large volumes. The upper limit is 10^{20} m^{-3}. The density of most magnetically confined plasmas is close to the limit. Hence, the important parameter is the time τ. A significant time should depend only upon physical properties of the system. The *energy confinement time* τ_E (see Box 5.1) is an adequate figure of merit for tokamaks.

Plasma confinement by magnetic field is far from perfect. Particles diffuse through field lines and are lost. Since they carry kinetic energy, some energy is lost accordingly. This loss adds to radiative losses. Altogether, relative energy leaks are faster than relative particle leaks. Between the particle confinement time τ_p and the energy confinement time τ_E, the latter is shorter. It can be evaluated experimentally by looking at the decay of the plasma temperature when all heating mechanisms are stopped.

In order to make simpler the comparison among different tokamaks, a single criterion is used which combines Lawson product and the temperature in the *triple product* of the particle density, the plasma temperature and the energy confinement time. The triple product is the ultimate factor of merit. It is expected to exceed a threshold value in a successful fusion device.

Box 5.1. Energy confinement

The energy confinement time τ_E is defined after the rate equation for the plasma internal energy W ($\propto$ nT), viz. $\frac{dW}{dt} = P_{Ch} - \frac{W}{\tau_E}$ where P_{Ch} represents the heating power. Energy losses are greater than or at least equal to bremsstrahlung losses. In a $\frac{nT}{\tau_E} \geq bn^2T^{1/2}$, a factor n cancels out and $n\tau_E$ should be smaller than a value proportional to $T^{1/2}$: *the bremsstrahmung limit.*

The goal of magnetic fusion is represented in compact form by the inequality

$$nT\tau_E > 3 \times 10^{21} \, m^{-3} \cdot keV \cdot s.$$

In future thermonuclear reactors, the confinement time is expected to be adjustable and, therefore, much longer than the energy confinement time. Particle and energy losses are to be compensated. Matter should be introduced in the plasma while the reactor is working. Now, the plasma temperature obtained due to ohmic heating is too low (typically 2 keV) for an efficient fusion reaction. Additional heating should be provided in order to reach the required temperatures. The corresponding power is to be included in the energy balance. Ideally, once the ignition is achieved, the energy deposition by α particles balances energy losses. Actually, ignition will not be necessary provided a steady regime in which plasma heating (α particle energy deposition plus additional energy) exactly compensates the losses is obtained. The gain of the device is then the ratio of the fusion power to the additional heating power. It is expected to be high for a duration much longer than τ_E.

Given the particle density n and the temperature T, it is important to know the variations of τ_E with respect to geometrical and electrical parameters. Such variations are given by semi-empirical scaling laws elaborated the same way as the similarity relationships in fluid mechanics. The formulas combine factors like major radius, minor radius at the plasma boundary, safety factor on the magnetic axis, particle density, current intensity, additional heating power, etc., with semi-empirically determined exponents (Box 5.2).

Table 5.1. The size of a tokamak reactor.

Major radius R_0	~ 10 m
Plasma minor radius a	~ 3.5 m
Current intensity I_p	20 to 30 MA

Many teams running a tokamak proposed scaling laws. The simplest and best-known laws are used to set up a tokamak project.

Box 5.2. Examples of scaling laws

In the early 1970s, TFR (Tokamak at Fontenay-aux-Roses [1]) was a successful device from which the following relationship was derived: $\tau_E \propto \sqrt{q} n_e a^2 R_0$, which quite nicely fitted the results obtained with the first- and second-generation tokamaks. The machines were comparatively small and were not fitted with important auxiliary heating.

Another simple scaling law stems from the recent evolution of tokamaks. It reads: $\tau_E \sim HR^2 I_p P_{tot}^{-0.75}$ where H is a numerical factor ranging from 1 to 2 according to the prevailing confinement mode.

Another useful scaling law was set up on different grounds. It states that the square of the current intensity is proportional to the triple product (Goldston, 1985 [2]):

$$I^2(\text{MA}) = 1.4 \times 10^{-19} \, n T \tau_E.$$

Remembering the requirement the triple product should fulfill (see Box 5.1), one concludes that the current in a reactor is to exceed 20 MA. Then, the reactor size is estimated after the other scaling laws (Table 5.1).

A tokamak reactor is necessarily a large device. Hundreds of tonnes of materials and equipment are to be assembled. The price tag is so high that at the end of the 20th century, no state or community of states was willing to invest in such a machine. Consequently, ITER is a less costly project whose size and performances are below the requirements for a reactor.

3. Auxiliary heating

During the linear leading edge of the current pulse, the plasma temperature rises thanks to ohmic heating. Now, according to all tokamak experiments conducted so far, the conductivity increases in such a way that the plasma is comparable to superconducting materials when the electron temperature exceeds 2 keV. Ohmic heating is no longer effective. Energy is to be fed into the plasma in order to reach higher temperatures.

Three heating technologies were successfully used in the development of tokamaks: adiabatic compression, radiofrequencies and fuel injection.

3.1. Adiabatic compression

Since magnetic field lines are frozen in a strongly magnetized plasma, a slow displacement of field lines drives a plasma motion. The plasma can thus be compressed when field lines get closer to each other, i.e. when the magnitude of the magnetic field increases. Simultaneous tailored growths of the plasma current and the toroidal field coils current are to be performed. The method proved effective in the early 1970s at Princeton: the small tokamak ATC (Adiabatic Toroidal Compression) was specially designed for that purpose. For compression, the plasma current varied from 80 to 200 KA, whilst the toroidal field was growing from 2 to 2.5 teslas. Consequently, the major radius and the minor radius shrank from 0.88 to 0.35 m and from 0.17 to 0.11 m, respectively. The volume was reduced by a factor of 6, the electron temperature rose from 1 to 2.5 keV and the ion temperature rose from 300 to 800 eV. Although it was possible to implement the method in more recent devices, it is no longer in use.

3.2. Radio frequencies (RF)

Some regimes of wave propagation in magnetized plasmas include resonances in which the group velocity, i.e. the energy propagation velocity, is close to zero. The energy transfer to the plasma is then obviously enhanced. The mechanism is even more interesting when energy is directly given to ions. Ion cyclotron resonance (Figure 5.2) and lower hybrid resonance make it possible.

Given the magnitude of confinement fields, megawatt-power generators matching the required frequencies were developed. Such powers are indeed necessary for an efficient heating. Antennas are displayed inside the vacuum chamber. They are fed by high-frequency circuitry or by waveguides and beams travel along directions that favour resonances.

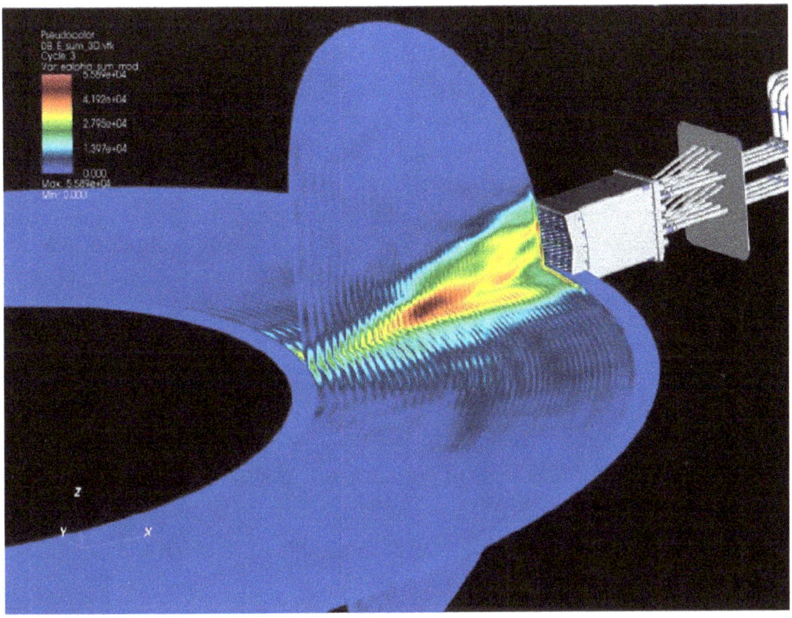

Figure 5.2. A simulation of the propagation of an ion cyclotron wave across a plasma ring. The beam emitted by the antenna is focused in order to get a maximum amplitude in the vicinity of the magnetic axis. At this location, the plasma density peaks. The simulation applies to ITER which is to be fitted with RF heating. This process proved effective in smaller tokamaks.

Now, RF beams have a useful spin off. Wave–particle interactions can be used to drive the plasma electrons so that an intense electric current is generated. This is an efficient way to relay the initial inductive pulse. The non-inductive *current drive* [3] adds up to the bootstrap effect. Altogether, the life span of the magnetic configuration can be extended. It exceeded 5 minutes (300 s) in Tore Supra, a third-generation tokamak run at CEA Cadarache, France. 400 s are expected for ITER.

The current drive can be spatially adjusted in order to shape the current profile across the torus so that the safety factor q is larger than unity everywhere in the plasma. By the same token, a *reverse shear* was obtained: q exceeds 2 on the magnetic axis, it first decreases as a function of the minor radius to a minimum greater than 1, and then increases on the way to the plasma boundary.

Box 5.3. Tore Supra and current drive

This tokamak, operating in France at CEA Cadarache, is smaller than JET and TFTR. However, it is considered as a third-generation device. Indeed, it started running in 1988 and brought significant technical advances. As the name indicates, the toroidal magnetic field is generated thanks to superconducting coils. Implemented full size for the first time, the technology proved effective and promising for a reactor in the future. Since there is no dissipation in a superconductor, direct currents can feed the coils for any duration. This property made Tore Supra an ideal test bench for current drive: in 2003, confinement times longer than 6 minutes were achieved [4].

3.3. Fuel injection

Every magnetic configuration undergoes particle leaks that can be compensated in two ways. First, solid DT pellets are sent through the plasma by a kind of gun. The hot plasma interacts with the pellet which is vaporized. This effect can be enhanced by a conveniently focused laser beam. The vapour expands, is ionized, and its particles eventually occupy the whole plasma, thus contributing to the conservation of the density.

In an alternative approach, injection of neutrals is used to compensate both particle leaks and energy losses. This is an effective heating process, provided neutrals have a kinetic energy, 100 keV to 1 MeV, much larger than the thermal energy of the medium they go through. Now first, only charged objects can be accelerated by electrostatic fields, and second, only neutrals can penetrate a magnetically confined plasma. Therefore, a neutral-beam injector is made of an ion source followed by accelerating grids, then, a neutralizing stage and finally, a magnetic device to eliminate remaining ions (Figure 5.3). Neutrals are ionized as soon as they enter the plasma, and the ions transfer their energy thanks to Coulomb collisions.

JET (Joint European Tokamak, running at Culham, UK, since the 1980s) is fitted with two sets of eight converging neutral injection lines (Figure 5.4). Each set is about 10 m high. Before neutralization, the ion energy exceeds 100 keV. The total power is 35 MW. In most experiment, deuterium is injected into the plasma. For the first experiments with tritium (1991), this isotope was introduced via the ion source in a single injection line. It was a safer way of coping with the radioactivity of tritium.

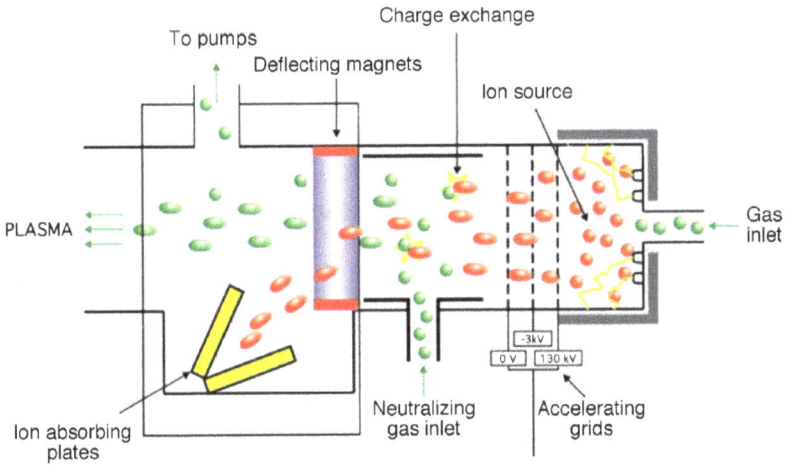

Figure 5.3. The principle of neutral injection in JET. Neutrals (in green) are created via charge-exchange collisions. To this end, accelerated positive ions (in red) are sent across a gaseous atmosphere. Charge-exchange collisions do not alter their energy and direction much. Magnetic fields deviate residual ions that eventually impinge onto absorbing plates.

Figure 5.4. An assembly of ion sources for neutral injection (JET).

1 MeV neutral beams would deliver a very efficient heating. Due to decreasing charge-exchange cross sections, positive ions cannot be used with energy higher than 500 keV. Using negative ions would be free from this drawback. However, getting high intensities of 1-MeV negative ions is a real challenge [5]. Part of the ITER project is a research and development program on negative ion beams carried out mainly in France (CEA Cadarache) and Japan.

Implementing an auxiliary heating acts on the dynamics of a tokamak. In first-generation devices, a slow, approximately linear growth of the electron temperature was unravelled by X-ray diagnostics. The 20% increase ends with an abrupt decay back to the initial temperature, and the cycle resumes: successive *sawteeth* are thus observed. The temporal width of the saw tooth increases with the size of the tokamak. It was slightly shorter than 2 milliseconds in TFR, a state-of-the-art machine of the early 1970s (Figure 5.5). It is about 100 milliseconds in JET. The decay occurs in less than 100 microseconds.

As it is nowadays firmly established, the sawtooth collapse corresponds to major changes in the magnetic map, islands forming and growing at the flux surface q = 1 (see Chapter 4).

Sawteeth were commonplace in first-generation tokamaks. They no longer appear in more efficient confinement regimes obtained with bigger machines. The magnetic configuration tolerates some slowly growing instabilities that lead to magnetic islands. However, magnetic reconnection induces partly chaotic changes in the magnetic topology. Disordered field lines contribute to an enhanced particle transport across the magnetic field. Instabilities and anomalous transport are still major problems of tokamak research.

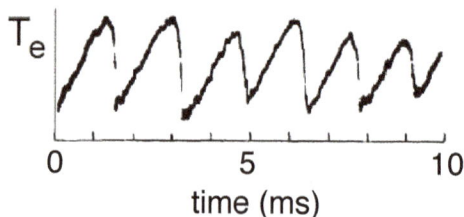

Figure 5.5. The saw teeth observed at TFR (1974) in the time history of the electron temperature. Time in milliseconds (ms).

4. Three generations of tokamaks

The main data that describe machines built since 1970 are listed in Table 5.2. They are classified according to the generation devices they belong to. Major trends are noteworthy: increasing size, D-shaped meridian cross sections and implementation of divertors.

Table 5.2. Successive generations of tokamaks: current and volume increase as a generation replaces the previous one. However, the toroidal magnetic field is not very different from one device to another. The upper limit is imposed by the strength of the materials the coils are made of.

		I (MA)	R_0(m)	a(m)	B_T (T)
The ancestor	1960s				
T3 (Moscow)	USSR	0.09	1	0.17	2.5
First generation	1970–75				
ORMAK (Oak Ridge)	USA	0.2	0.8	0.23	1.8
ST (Princeton)	USA	0.13	1.09	0.13	4
ATC (Princeton)	USA	0.08; 0.2*	0.88; 0.35*	0.17; 0.11*	2; 2.5*
TFR (C.E.A.)	France	0.4	0.98	0.2	6
ALCATOR (M.I.T.)	USA	0.16	0.54	0.095	6
Second generation	1975–80				
DITE (Culham)	GB	0.28	1.12	0.23	2.8
PLT (Princeton)	USA	1.4	1.3	0.45	4.6
FT (Frascati)	Italy	1	0.83	0.21	10
T 10 (Moscow)	URSS	0.8	1.5	0.35	5
ASDEX (Garching)	Germany	0.5	1.54	0.4	3
DOUBLET 3 (San Diego)	USA	2**	1.67	0.45**	2.6**
Third generation	1980–90				
DIII-D (San Diego)	USA	3.5	1.67	0.67; 1.3***	4.2
JET (Culham)	Great Britain	5	2.96	1.25; 2.1***	3.5
TFTR (Princeton)	USA	2.5	2.48	0.85	5.2
JT 60 (JAERI)	Japan	3	3	1	5
TORE SUPRA (Cadarache)	France	2	2.25	0.7	4.5
Fourth generation	2019				
ITER (Cadarache)	International	*15*	*6*	*2/3.2*****	*5.3*

*First value: uncompressed plasma; second value: after adiabatic compression.
**Values for a single ring.
***The 2 values of the minor radius account for a D-shaped meridian cross section.

4.1. First generation

It appeared during the early 1970s. The devices are comparable to the successful T-3 in terms of their sizes and performances. Two among them were remarkable by their high magnetic field: ALCATOR (Massachusetts Institute of Technology, USA) and TFR* (Fontenay aux Roses, France). The latter yielding outstanding results like the discovery of saw teeth and the derivation of a scaling law which served as a reference for quite a long time.

4.2. Second generation

Two devices in this generation are worth mentioning: ASDEX still running in Germany (Garching) and Doublet III in General Atomics laboratories (San Diego, USA). The latter displayed a special design: the vacuum chamber could accommodate two superimposed plasma rings, hence the name. Various meridian cross sections could thus be tested. Actually, experimentalists scarcely used that possibility. Doublet III was converted into a third generation machine and renamed DIII-D [6] with a D-shaped meridian cross section as in JET.

ASDEX and Doublet III are famous in the history of magnetic fusion. Indeed, in these devices, a high confinement (H) mode was evidenced [7]. As the plasma was exposed to the intense heating by neutral beams, the confinement suddenly improved, and the turbulences at the plasma edge all but disappeared. In the H mode, the confinement time is twice as long as in the usual confinement mode of first-generation tokamaks or L mode (for low confinement). The H mode shows up in a tokamak whose field map includes a separatrix when the auxillary heating power increases. The net power (radiation-induced) crossing the last closed flux surface has to exceed a machine dependent threshold. The transition is made easier if the internal wall of the vessel is covered with "tiles" made of light elements, carbon or beryllium. Losses to radiating heavy impurities are thus minimized.

The H mode is a first step towards an *advanced confinement mode* (Figure 5.6) in which peak plasma temperature and density are even higher with different spatial profiles due to the occurrence of an internal transport barrier.

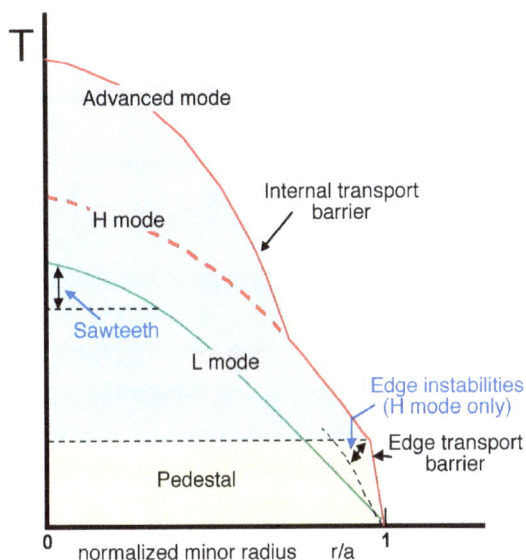

Figure 5.6. The temperature profiles along a minor radius in different confinement modes. In the L mode, small tokamaks undergoes instabilities that generate sawteeth. In the H mode, the temperature is higher, and an unstable transport barrier appears in the vicinity of the plasma boundary. In the advanced confinement mode, another transport barrier, this one stable, takes place half way from the magnetic axis to the plasma boundary.

Table 5.3. History of a third-generation tokamak: JET.

1973–75	Design under the supervision of P. H. Rebut
1975–77	International discussions for the choice of the host country
1983	First plasma
1991	First experiment with a D–T mixture
1997	P_{fusion}: 16 MW (D–T), Energy per shot: 13 MJ.
2000	Management is transferred to Great Britain within an Euratom association.
2010	Still running: experiments aim at preparing ITER

4.3. Third generation

Building and running big tokamaks is costly, hence the small number of third-generation machines: JET [8] and Tore Supra [9] in Europe, DIII-D and TFTR [10] in the USA, JT-60 in Japan. The time necessary for the design and the construction was quite long: 7 years in the case of TFTR, 10 for JET, a joint venture of the European community (Table 5.3). Indeed, such projects are exceedingly complex in all domains: science, technology and administration.

JET was an innovative machine with a D-shaped meridian cross section. In the new Japanese device JT-60, designed to contribute to the ITER project, the meridian cross section is also D shaped. Plasma performances are close to those obtained at JET. However, tritium fuelling is excluded.

Compared to JET, TFTR (Tokamak Fusion Test Reactor built at Princeton, USA, final shot in 1997) was a simpler device with a circular meridian cross section. The current did not exceed 3 MA. The project incorporated a two-component plasma [11]. The ion distribution has a strengthened high-energy tail due to neutral injection and controlled slowing down of the subsequent ions. Furthermore, in final runs, the reverse shear was tested and actually did improve the confinement. In normal shear, the field vector rotates clockwise along a minor radius, starting from the magnetic axis. Shear reversal (counter-clockwise rotation) at small minor radii occurs provided the current profile is adequately tailored by current drive. Then, the safety factor is larger than unity within the whole plasma ring.

Plasmas made of D–T mixtures were used at TFTR and JET. In 1993, 10 MW of D–T fusion power were obtained at TFTR. At JET in 1997, the yield was 16 MW of fusion power versus 21 MW of auxiliary heating power, i.e. a gain of $Q = 0.65$. However, these results correspond to conditions (very high temperature for a short time) that are far from the expected working point of a reactor (Figure 5.7).

Using tritium implies hazards due to radioactivity. Tritium itself is radioactive with a half-life of 12 years. Furthermore, the 14 MeV neutrons from the D–T reaction induce radioactivity in all structural materials. Therefore, in tokamak projects, runs with tritium are to be conducted cautiously. They are planned near the end of experimental campaigns. At TFTR, the first plasma was obtained in December 1982, and the first D–T reactions in 1994, 3 years before the last run in April 1997.

5. Assessing tokamaks

Magnetically confined reactors are compared on a logarithmic plot: in abscissas, the plasma temperature; in ordinates, the triple product

Fusion power

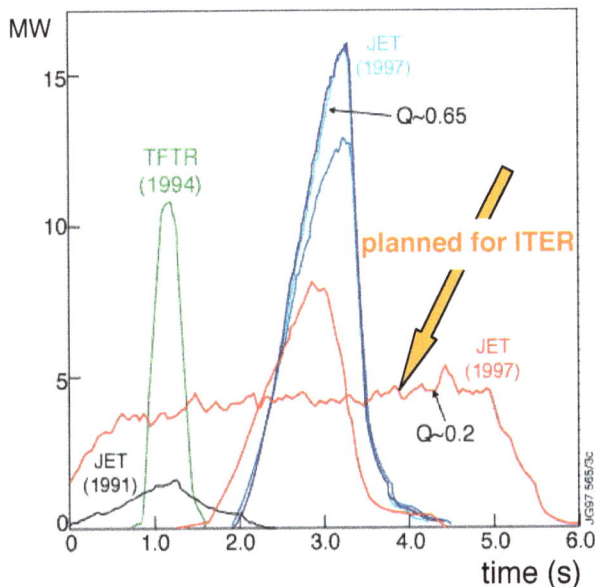

Figure 5.7. Experiments with tritium in JET and TFTR: the time history of D–T fusion power during various experiments. Sharp peaks correspond to conditions far from reactor operation. The expected regime in ITER (400 MW during 400s) is the same as the 4 MW fusion power for the whole duration of the confinement obtained in JET. The gain will be 10 instead of 0.2.

(Figure 5.8). The *break-even* is obtained when the fusion power equals the auxiliary heating power. This corresponds on the diagram to a curve that passes through a minimum triple product slightly below 10^{21} m^{-3}·keV·s at temperature 15 keV. At this temperature, ignition or a least a high gain requires a triple product over 5×10^{21} m^{-3}·keV·s. Such conditions are required for a possible energy source in the future.

This evolution is indeed an impressive one, the more so when compared to Moore's law, stating that the number of transistors on a computer chip doubles every two years. Actually from T-3 to JET, the triple product of tokamak has doubled every 1.8 years (Figure 5.9).

As time elapsed, the fusion power from tokamak increased significantly. D–T runs were performed only at JET and TFTR. For other devices, data

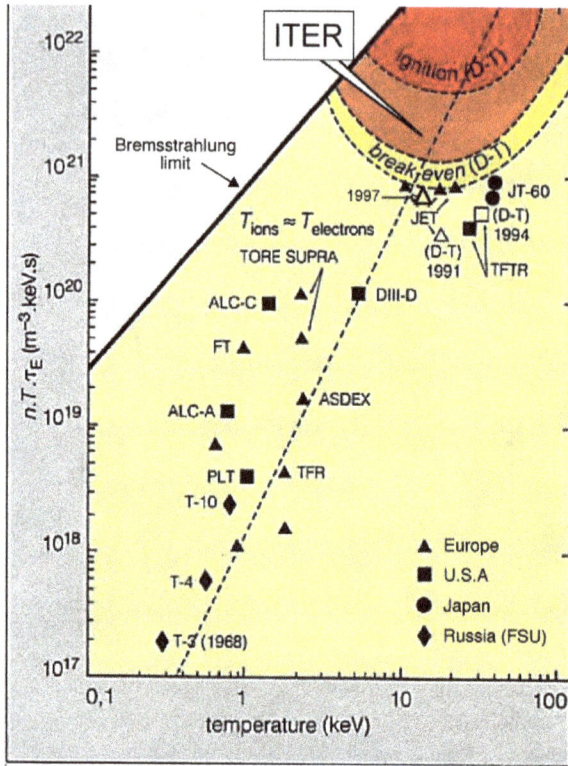

Figure 5.8. 30 years of tokamaks. From 1968 to 1997, the triple product increased by 4 orders magnitude, whilst plasma temperature was growing by a factor of 20. ITER is expected to be well over the "break-even" limit: fusion power equals heating power. Third-generation tokamaks came close to the break-even conditions but remained still below.

are available after D–D runs from which an equivalent D–T power can be inferred. The time history of the fusion power turns out to be similar to the time history of the triple product.

Implemented during the 1980s, third-generation tokamaks were fully operational in the 1990s. Many results which are of interest for the next step, i.e. the ITER project, were obtained. Note that auxiliary heating has proved necessary during the whole confinement time. This changed the perspective. Ignition is no longer considered as mandatory. The goal of future tokamaks is a high gain, i.e. a high ratio between the fusion power and the heating power.

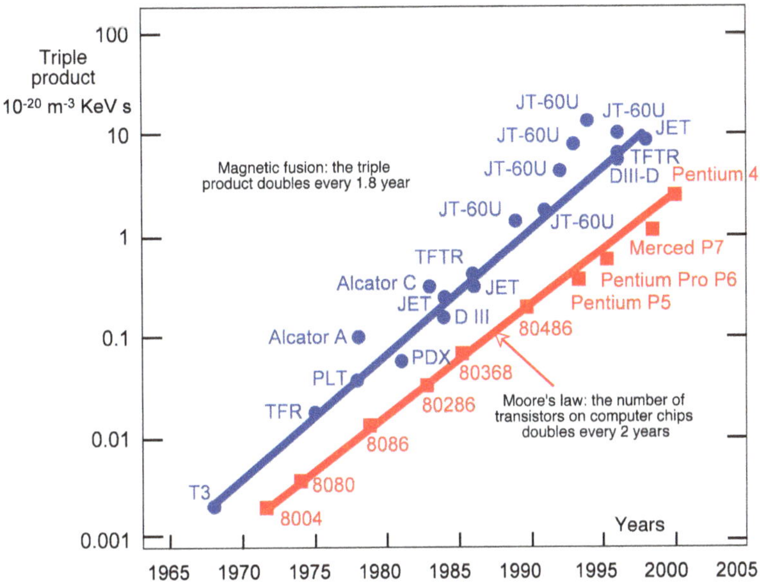

Figure 5.9. Compared time histories of the number of transistors on a computer chip and of the triple product in tokamaks. The former increased according to Moore's law, doubling every 24 months. The latter has doubled every 22 months. In ITER, the triple product aimed at is about 30×10^{21} m^{-3}·keV·s.

Box 5.4. Fusion and the European community

Since the late 1950s, controlled thermonuclear fusion is a major project in the R&D portfolio of the European community. As a part of Euratom programs, research is conducted via associations with laboratories in member states. The community makes available subsidies and manpower. Scientists with Euratom status join the project provided they are not working in the country they are citizen of. This policy proved effective to strengthen international cooperation within Europe. On the negative side, decision-making processes are slow and cumbersome.

The energy crisis following the oil shock of 1973 was an opportunity for a project that the community could directly manage towards an electrical energy source. This is the origin of the JET (Joint European Torus), a device archetypal of third-generation tokamaks. As shown in Table 5.3, the total duration of the project spans many decades: 24 years from the first sketches to significant results with D–T mixtures.

JET was a fully European laboratory till January 1, 2000. Afterwards, the device was run by Great Britain (Culham research centre) within the frame of ordinary Euratom contracts.

The community through Euratom is a major partner in the ITER project. It will provide 46% of the building cost and 34% of the running cost.

References

1. TFR group, *Plasma Physics and Controlled Fusion* **19** (1977) 349; *J. Phys. Colloques* **38** (1977) C3-21-C3-32.
2. R. J. Goldston, *Plasma Physics and Controlled Fusion* **26** (1984) 87.
3. N. J. Fisch and C. F. F. Karney, *Rev. Mod. Phys.* **59** (1987) 175.
4. D. van Houtte *et al.*, *Nucl. Fusion* **44** (2004) L11.
5. Special issue on Negative ion based neutral beam injection, *Nuclear Fusion* **46** (2006).
6. Holzhauer *et al.*, *Plasma Phys. Control. Fusion* **36** (1994) A3.
7. F. Wagner *et al.*, *Phys. Rev. Lett.* **49** (1982) 1408.
8. http://www.jet.efda.org/.
9. http://www-fusion-magnetique.cea.fr/gb/.
10. https://fusion.gat.com/global/DIII-D.
11. http://www.pppl.gov/projects/pages/tftr.html.
12. J. M. Dawson, H. Furth and F. H. Tenney, *Phys. Rev. Lett.* **26** (1971) 1156.

6

ITER and Satellite Programs

1. A unifying project

Since fusion devices grow bigger and so does their cost, decision making needs more time and construction takes more time. As in space exploration, planning for the next step when the present one has just gotten underway is a necessity. Fusion scientists thus have ample time to prepare the next generation. Starting mid 1970s, whilst third-generation machines were being built, teams in the FSU, in Japan, in the USA (INTOR for INternational TOkamak Reactor) and in Europe (NEXT for NEXt European Tokamak) independently designed tokamaks aimed at ignition.

Summits between the US president Ronald Reagan and the first secretary Mikhail Gorbatchev of Soviet Union took place in 1985 and 1986. Both heads of states were looking for a common project that, through scientific and technical cooperation, could exemplify improved relationships between Eastern and Western countries. Furthermore, both scientific advisors, Alvin Trivelpiece for the USA and Evgenyi Velikhov for the Soviet Union, had made a reputation in the field of plasma physics. Magnetic fusion appeared as an ideal endeavour. Teams were already working on new generation tokamaks beyond JET and TFTR [2]. Institutionalization of a single project involving all these teams was then decided. So was the inception of the ITER project whose existence was not impaired by the collapse of the USSR.

Initially ITER aimed at ignition, hopefully obtained during the 2010s. The achievement would be comparable to the first fission chain reaction experiment, which made history on December 2nd, 1942, at Chicago under Enrico Fermi's leadership.

The project phase took some 15 years. In July 1998, the design of a prototype reactor (ITER-EDA for Engineering Design Activities) was completed on schedule. The reactor was to achieve controlled ignition (1.5 GW for 1000 s) and demonstrate tritium breeding. However ignition was still uncertain. An estimate of the construction cost was 7 billions dollars. As for other big science endeavours of the 1990s, the project was rated too expensive for a questionable result. A downsizing was then decided: build a 1.3 times smaller machine at "half price".

In the new project (ITER-FEAT for Fusion Energy Advanced Tokamak), ignition is given up although it is not completely excluded. The reduced technical objectives are: to achieve extended burn in inductively driven plasmas at $Q > 10$ for a range of scenarios, whilst not precluding the possibility of controlled ignition; to aim at demonstrating steady state operation through current drive at $Q > 5$. Both ITER projects were grounded on data from second- and third-generation devices. After these results, a scaling law in H mode combining eight engineering parameters was set up (Table 6.1).

Table 6.1. Parameters for a semi-empirical scaling law in H mode.

I_p	Total current intensity through a meridian cross section
B_T	Toroidal magnetic field at the magnetic axis
P_{tot}	Total power absorbed by the plasma
$n_L = \frac{n_e V}{2\pi R_0}$	Linear average density in a plasma with volume V
R_0	Major radius of the torus
$\varepsilon_A = \frac{a}{R}$	Reciprocal of the aspect ratio
$K_A = \frac{V}{2\pi^2 R_0 a^2}$	Average elongation expressed, in volume ratio, of the non-circular meridian cross section of the plasma
M_{eff}	Average mass number of the plasma ions (2.5 for a D–T mixture)

Scaling law: $\tau_E = 0.079 I_p^{0.89} B_T^{0.21} P_{tot}^{-0.49} n_L^{0.11} R_0^{1.58} \varepsilon_A^{0.23} K_A^{0.52} M_{eff}^{0.5}$

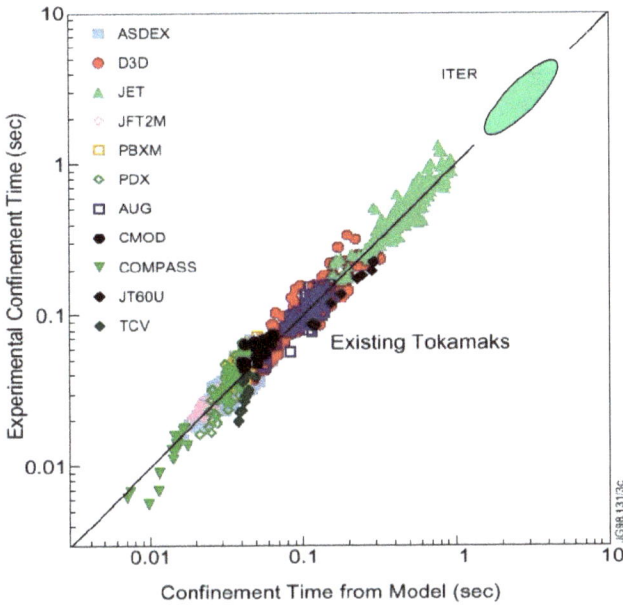

Figure 6.1. Comparing a scaling law to experiment. On a logarithmic plot: abscissas are energy confinement times according to "Model", i.e. the scaling law of Table 6.1 for a number of tokamaks in H mode; ordinates are the corresponding experimental data. All points are close to a line with slope 1, meaning approximate equality. Extrapolating the scaling law to ITER, the expected energy confinement time ranges from 1.8 to 5 s.

The relevance of the scaling law is readily seen on Figure 6.1. It is a logarithmic plot in which ordinates are experimentally measured energy confinement times from many tokamaks and abscissas are the corresponding values from the scaling law. All points lie in the vicinity of the line with slope 1. With a reactor size tokamak (see Table 5.2) or the slightly smaller actual ITER, the energy confinement time is expected to range from 1.8 to 5 s. Table 6.2 compares 2001 ITER to 1998 ITER and JET.

ITER is an international project seven parties are committed to: the European community (via Euratom), China, South Korea, United States, India, Japan and Russia. Europe contributes 40% of the total budget (construction of the tokamak plus running during 25 years). France contributes one-fifth of European expenditures. Every other party contributes 10%. Since the international partners signed up to the project in 2006, the

Table 6.2. From JET to ITER (final project — 2001). Expected performances of 2001 ITER are well below ambitious 1998 ITER.

	JET (1975)	ITER 1998	ITER 2001
Major radius of the plasma ring (m)	2.96	8.1	6.2
Plasma half width in the equatorial plane (m)	1.25	2.8	2.0
Toroidal field at the magnetic axis (T)	3.5	5.6	5.3
Rated plasma current (MA)	5	21	15
Neutral injections power (MW)	25	50	40
Rated fusion power (MW)	4	1500	$\varepsilon 400$
Gain Q $(= P_{fusion}/P_{heating})$ (standard plasma)	0.2 − 0.7	∞	$\varepsilon 10$
Rated duration of the current plateau (s)	5	>1000	>400
Average neutron flux density at first wall (MW/m^2)	0.05	∼1.0	0.57
Initial projected costs Construction	—	$7 \cdot 10^9$	$3.5 \cdot 10^9$
Running (25 years)	—	$3 \cdot 10^9$	$1.5 \cdot 10^9$

total cost including running has increased fourfold to around €15 billion (US$19.4 billion), and the original date of completion was shifted by four years.

In 2003, it was decided to build and run the 2001 ITER. Two year later, the site was chosen next to the Cadarache nuclear centre in southern France (Figure 6.2). Ground breaking occurred in 2007. Building of a 40-hectares platform ended in 2009. Implementation of antiseismic supporting pillars was completed at the end of 2012. So was the construction of office buildings. The construction tasks for the tokamak itself are split into subsets distributed among parties (Figure 6.3).

2. How ITER looks like

As shown on Figure 6.3, ITER has a D-shaped meridian cross section. Three major components are: superconducting coils, a divertor and a *blanket*. Superconducting coils were satisfactorily implemented at Tore Supra. Divertors were also successful at JET and JT-60. The blanket which converts 14 MeV neutrons' energy into heat will be tested for the first time.

The blanket will be made of removable modules (Figure 6.4). Every module is 450 mm thick and incorporates the first wall facing the plasma.

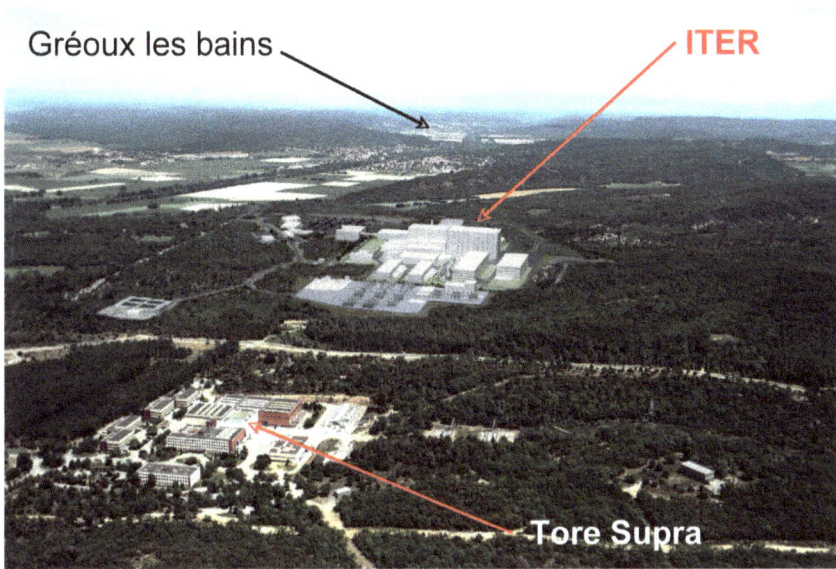

Figure 6.2. ITER is located in France, next to Cadarache nuclear research centre, at walking distance from the third-generation tokamak Tore Supra. Bulldozer started clearing ground in 2007. The first building was dedicated in January 2013.

The requirements for the first wall are: transparency with respect to neutrons, resistance to thermal stress and to plasma–wall interactions, mechanical stiffness in operating conditions. The energy flux at the first wall is 200 to 500 kW/m^2.

The bulk of the module is to absorb the 14 MeV neutron flux whose power is 600 to 800 kW/m^2. The neutron energy is converted into heat to be evacuated with a liquid coolant, in the ITER case, water whose temperature will not exceed 200°C under a 30 bar pressure. Such conditions preclude an efficient powering of electrical generators.

Some blanket modules will be devoted to tritium breeding experiments. Fission of Lithium nuclei (see Chapter 2) will be implemented. Lithium will be incorporated either in ceramics or in an LiPb eutectic.

Finally, the blanket should protect the surroundings. Indeed, superconducting coils, structural elements, control and measuring instrumentation are to be shielded from all kinds of radiation emitted by the plasma: neutrons, charged particles, electromagnetic waves.

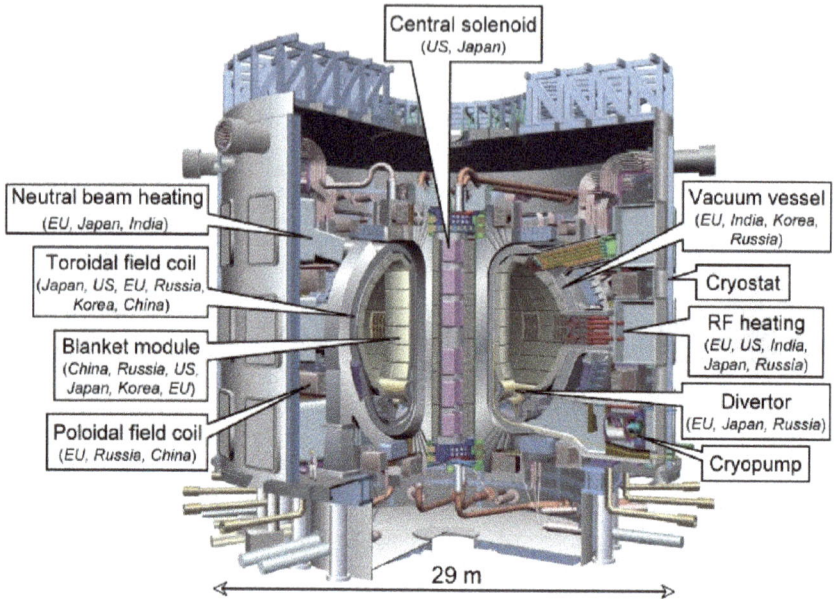

Central solenoid
(US, Japan)

Neutral beam heating
(EU, Japan, India)

Toroidal field coil
(Japan, US, EU, Russia, Korea, China)

Blanket module
(China, Russia, US, Japan, Korea, EU)

Poloidal field coil
(EU, Russia, China)

Vacuum vessel
(EU, India, Korea, Russia)

Cryostat

RF heating
(EU, US, India, Japan, Russia)

Divertor
(EU, Japan, Russia)

Cryopump

29 m

Figure 6.3. An artist's view: cutaway of ITER. The whole machine is nested inside a cylinder with height and diameter both equal to 29 m and kept at low temperature (cryostat). Toroidal superconducting coils are immersed in liquid helium. The meridian cross section is D-shaped.

Actually, ITER will be part of a larger program intended at a magnetic confinement fusion reactor. As construction proceeds, JET and Tore Supra are still running. Furthermore, since 2006, a joint project of the European community and Japan aims at upgrading the Japanese tokamak JT-60. In the new device called JT-60SA (for Super Advanced) [4], superconducting coils will be used to create the toroidal magnetic field. The plasma ring will be of JET class. It will look like a downsized ITER with a current ranging from 3.5 to 5.5 MA. The first plasma is scheduled for 2015. JT-60SA is to serve as a test bench for technologies beyond ITER. Another auxiliary program dealing with material testing in reactor-like conditions will be described in Section 6.4 and 6.5.

ITER itself will be a multipurpose test bench that will serve to prepare a more decisive step. Its objectives are both physical and technological.

From the physical view point: producing and assessing a plasma heated mainly by α particle energy deposition; achieving for 400 seconds a fusion

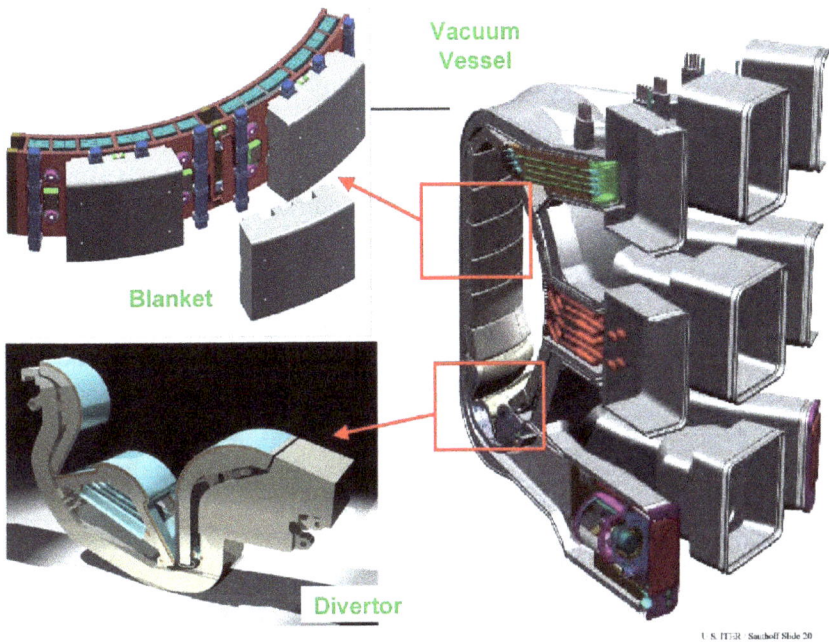

Vacuum Vessel

Blanket

Divertor

U. S. ITER Sauthoff Slide 20

Removable blanket modules and the divertor in ITER. The modules are hooked to the inner wall of the vacuum vessel. In the divertor, particle absorbing areas are covered with tungsten or carbon layers.

power 10 times higher than the external heating power (α particle energy deposition is then twice the heating power); achieving a steady regime with a fusion power 5 times the heating power; investigating conditions for ignition.

On the engineering side: full-sized demonstration of technologies to be implemented in a fusion power plant; testing first walls that are to withstand energy density fluxes and fluences of order $0.5\,MW/m^2$ and $0.3\,MW{\cdot}year/m^2$ respectively; testing tritium regeneration.

ITER is managed with extreme caution. According to the initial schedule, the first plasma was due by 2015 and runs with tritium were to start by 2020. Now it appeared to the steering committee that engineering problems induced displacements of the milestones and a severe increase of the costs. Having pieces produced in different countries, then shipped to ITER building site and assembled there did not prove as efficient as expected. Consequently, the date of first plasma has slipped first to 2018 then to late 2020. The energy producing experiments will not come before 2026, nearly two decades after ground breaking.

3. ITER and safety

As a nuclear plant, ITER is to fulfill safety requirements. However, problems associated with radioactivity are far less stringent than in the case of fission power plants. Since there is no fissile material, running a fusion reactor does not produce nuclear waste in the usual sense: neither fission products nor actinides. There is still some radioactivity left. Indeed, 14 MeV neutrons activate the blanket and structural materials. Maintenance or repair will be carried on the same way as at JET. Most tasks are to be performed by remote handling. Workers in radiation suits will be occasionally allowed to penetrate inside the vacuum chamber.

When running of ITER comes to an end, the radiation level will depend upon the nature of structural materials. For instance, with vanadium steel facing the plasma, it will be, at shutdown, 10 times smaller than in the case of a PWR (pressurized water fission reactor) vessel. Since activation creates mainly short-lived isotopes, the residual radioactivity level will be negligible after a few decades (Figure 6.5).

Cadarache is classified as an area of moderate seismic activity; accordingly the tokamak will rest upon pillars designed to withstand earthquakes. ITER facility will be equipped with seismic sensors around the site to record all seismic activity, albeit minor.

Two other concerns are related to ITER safety. First, fearing an explosive runaway of the fusion reaction can be considered as pure fantasy. Indeed the plasma ring accommodates less than 1 gram of D–T fuel inside 800 cubic meters. Should anything happen, the plasma cools instantly. The nuclear reaction is immediately quenched.

A more serious worry is tritium, a β-radioactive isotope with half-life 12 years. Several kilograms will be present on-site, divided in lesser quantities. Based on feedback from JET and from other research laboratories, modern and efficient safety measures for the storage and handling of tritium have been incorporated into the ITER design. The facility will be protected against tritium release. Small quantities that might accidentally leak would be quickly diluted in the atmosphere (tritium is a

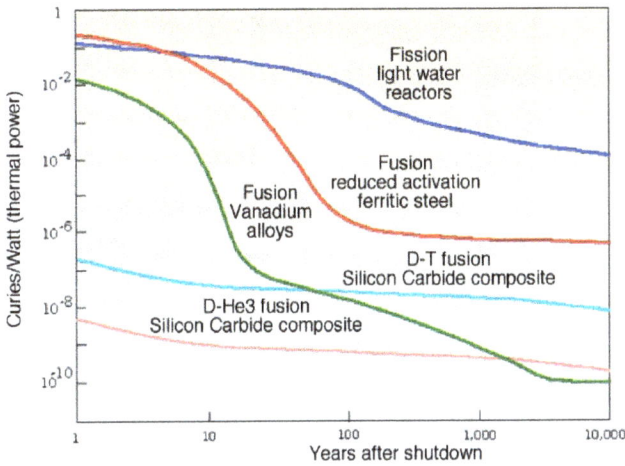

Figure 6.5. Time histories of radioactivity levels after shutdown for different first wall materials and fusion fuels. For comparison, the curve in blue is the time history of the vessel radioactivity after the shutdown of a LWR (light water reactor) fission plant.

hydrogen isotope still much lighter than air, recall) and carried up to the upper atmosphere.

4. The question of materials

With ITER and accompanying fusion devices, all scientific and technical aspects of tokamaks are to be tested but one: the strength of materials, especially the first wall, that are to withstand high heat and 14 MeV neutron fluxes in reactor conditions. The first project, 1998 ITER, did not overlook the problem. Preliminary studies on the strength of materials were to be carried on during the planned long construction time.

In a fusion reactor, energy (heat) fluxes are among the highest in technological devices, as shown in Figure 6.6 [5]. Depending upon which part is considered they are greater than or just comparable with fluxes in fission reactors. In routine conditions, they are surpassed only by fluxes encountered in space technologies: rocket nozzles and re-entry shields. In accidental circumstances, fortunately scarce for major disruptions, for short times, heat fluxes in structural materials can reach values that might end in catastrophic damages.

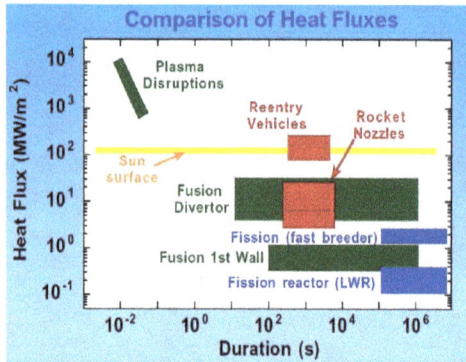

Figure 6.6. Heat fluxes that impact materials in space (in red) rockets and vessels or in nuclear fission (in blue) devices are plotted versus duration (in seconds) on logarithmic scales. LWR stands for light water reactors. In green are the conditions expected for important parts of ITER. In the diagram, the upper duration limit for fission reactors (about a year, i.e. 3×10^7 s) is beyond the range of abscissas.

Radiative effects on materials are of two kinds. All types of radiation: EM waves, charged particles, neutrons, carry energy. Materials are to withstand huge energy fluxes. In addition, neutrons displace atoms and trigger nuclear transmutations. Hydrogen or helium atoms are thus released and coalesce into bubbles along dislocations and grain boundaries, eventually inducing major changes in the mechanical properties (Figure 6.7).

Such changes are temperature dependant with different thresholds according to the nature of the material. For instance, in the case of steel,

- Below 400°C, reduced ductility, brittleness, smaller resistance to fracture, etc. are expected.
- Between 300 and 600°C, a swelling might occur, e.g. RAFM steel has a maximum near 450°C (see Box 6.2).
- Above 600°C, creep and precipitation increase dangerously.

Material irradiation by neutrons is measured by the average number of displacements, dpa, an atom undergoes during a given time (see Box 6.1). In addition, nuclear reactions occur which generate gaseous elements. Hydrogen or helium creation rates are counted in appm, i.e. atoms produced per million displacements. Table 6.3 presents expected data for ITER, and in the future DEMO and a commercial reactor. In all cases, the transmutation

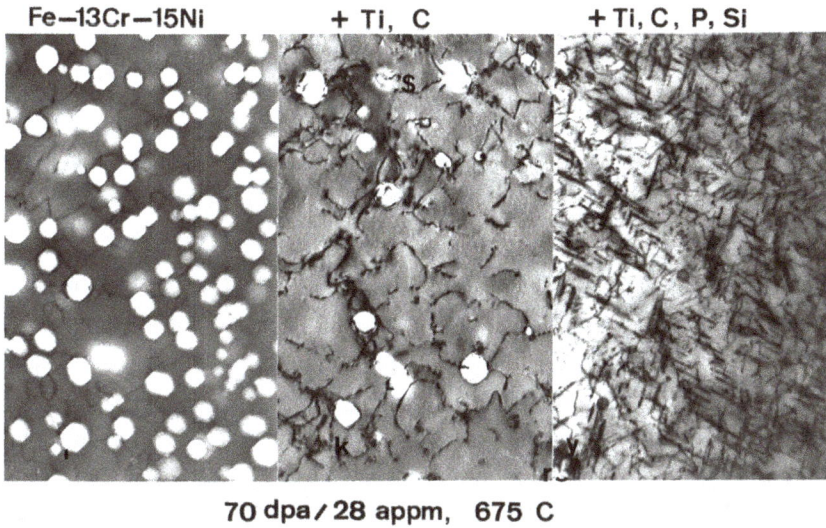

Fe–13Cr–15Ni | + Ti, C | + Ti, C, P, Si

70 dpa / 28 appm, 675 C

Figure 6.7. Consequences of irradiation for various steels. Under identical irradiation, a standard stainless steel is full of bubbles; added titanium and carbon considerably reduce the number of bubbles; bubbles disappear completely whenever the steel also contains phosphorus and silicon. However, in the latest case more dislocations and stacking faults are observed. Electron micrographs by L. K. Mansur and E. H. Lee (BNL) [6].

rates are much higher than 0.2 to 0.3 appm He/dpa, typical figures for fission reactors.

Box 6.1. Displacements per atom

In matter, neutrons slow down via collisions with atoms. Each time a neutron knocks an atom part of its momentum is transferred. The atom is slightly displaced from its initial location to which it will come back thanks to the high temperature due to absorbed radiative energy. Let a flux ϕ impinge onto N_0 atoms. The number of displacements is $N_D = N_0 \phi \sigma_D t$, where σ_D is the cross section for a single displacement. The number of displacements per atom (dpa) is the ratio N_D/N_0. The reference time is usually a year.

ITER is going to be a strong neutron emitter. However, the flux density will not exceed 10^{14} neutron/cm^{2}s. Similar values are encountered in high flux fission reactors such as ILL (Institut Laue Langevin in Grenoble, France). Only neutron energy spectra are different: 14 MeV at ITER, thermal at ILL.

Table 6.3. Irradiation conditions in ITER, DEMO and in a future commercial reactor.

	ITER	DEMO	Reactor
Fusion power	0.5 GW	2–2.5 GW	3–4 GW
Thermal flux (first wall)	0.1–0.3 MW/m^2	0.5 MW/m^2	0.5 MW/m^2
Neutron energy flux (first wall)	0.78 MW/m^2	< 2 MW/m^2	~ 2 MW/m^2
Fluence (first wall)	0.07 MW/m^2 (3 years)	5–8 MW·yr/m^2	10–15 MW·yr/m^2
Displacement per atom	< 3 dpa	50–80 dpa	100–150 dpa
Transmutation rate (first wall) leading to helium or hydrogen bubbles	~ 10 appm He/dpa		~ 45 appm H/dpa

NB. dpa = number of displacements per atom; appm = atoms produced per million.

At the ITER level there is no need for materials with a low activation potential. The problem of materials is to arise beyond ITER, in reactors where the first wall is expected to withstand neutron fluxes of order 10^{15} neutron/cm^{2}s. With the knowledge accumulated in nuclear industry, materials are already known to be suited for a future fusion reactor (see Box 6.2). Although metallurgical damage rates is comparable with fission reactors, transmutation rates that lead to hydrogen and helium bubbles are much higher (Table 6.4).

Box 6.2. Materials candidates for a fusion reactor [7][8]

Materials are to be as refractory as possible. Furthermore, neutron induced activation should be minimized. The following are expected to be suited to fusion reactors: high-temperature melting point metals such as iron, chromium or vanadium; carbon or silicon ceramics. Titanium is good only in ITER with less stringent conditions.
Among candidates:

1. Ferritic/martensitic stainless steels with reduced activation rates, i.e. RAFM (reduced activation ferritic martensitic).
2. Oxide dispersion strengthened (ODS) RAFM.
3. Vanadium (V) alloys.
4. Refractory metals: chromium (Cr) and tungsten (W) as well as their alloys.
5. Carbon (C/C) and silicon/carbide (SiC/C or SiC/SiC) composite ceramics.

Through a first selection, allocation of materials to specific functions in a fusion reactor was made possible (Table 6.5).

Table 6.4. Expected irradiation damages in a fusion reactor. Comparison with a state-of-the-art fission reactor.

Induced defects in steel	Fusion reactor (3–4 GW)	Fission reactor
Damage rate (dpa/year)	20–30	20
Helium produced (appm/dpa)	10–15	1
Hydrogen produced (appm/dpa)	40–50	10

Table 6.5. Choice of materials for a reactor classified according to their function.

Function	First wall	Fertile blanket	Divertor
Neutron multiplier		Be	
Tritium breeder		Li, Li Pb eutectic, ceramic including Li	
Structural materials	RAFM steel, Vanadium alloy, SiC		ODS RAFM steel, Tungsten alloy
Tiles facing the plasma	Tungsten, W alloy, ODS RAFM steel		Tungsten, W alloy, W on SiC
Coolant		Water, helium, Li Pb eutectic	Water, helium

5. IFMIF (International Fusion Material Irradiation Facility)

Untill 2030, materials in tokamaks will not be exposed to neutron fluxes of the same order of magnitude as in future fusion reactors. Now, researches should be made on materials withstanding such constraints. To this end, a specific facility, IFMIF [9] (for International Fusion Material Irradiation Facility) will be constructed and run as a satellite of the ITER program. Since no suited thermonuclear reactions will be available, a beam target configuration will be implemented. Lithium will be irradiated by accelerated deuterons and react according to:

$$\,^2_1D + \,^7_3Li \rightarrow 2\,^4_2He + n$$

D^+ ions will be accelerated up to 40 MeV in order to create 14 MeV neutrons with an optimized efficiency. The required fluxes will be

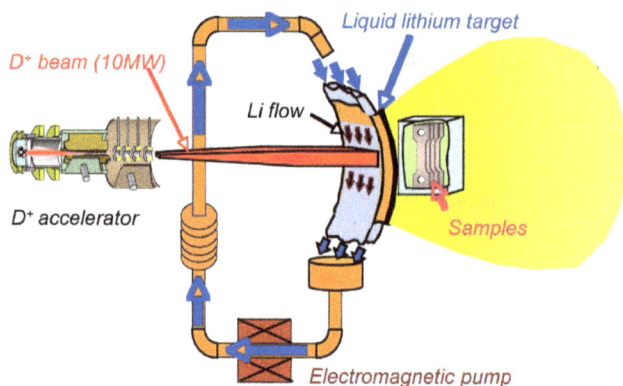

Figure 6.8. Scheme of IFMIF. The facility is a big science instrument. Altogether the lengths of the ion accelerator, the beam transport and focusing sections will exceed 100 m. A voluminous plumbing will be necessary to circulate liquid lithium.

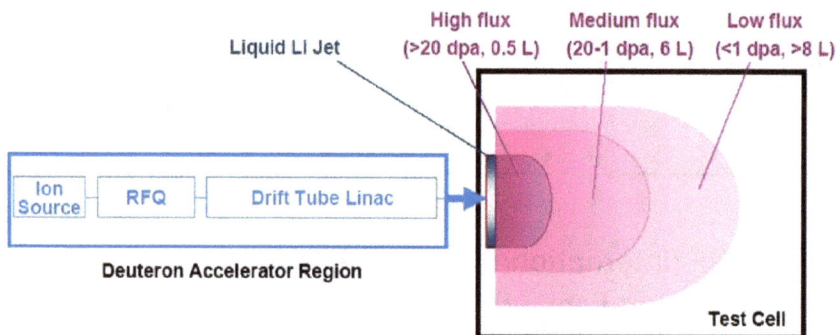

Figure 6.9. Irradiation map within the test cell. The available volumes (in litres) for fluxes larger than given values are given.

obtained using two 125 mA beams striking the same target. The necessary steady power is 10 MW. Ions will be focused onto a liquid lithium planar flow behind which samples to be tested will be irradiated (Figure 6.8 and 6.9).

IFMIF is a billion dollar international project. The Engineering Validation and Engineering Design Activities (EVEDA) phase, to be completed in 2017, is presently underway. In Japan, a prototype deuteron accelerator is being installed at Rokkasho to provide 125 mA of 9 MeV ions and a liquid lithium loop reaching 15 m/s of flow speed is running at Oarai in parallel with the development of its diagnostics and remote handling. The High Flux Test Module with a helium cooling loop and heaters capable of controlling the irradiation temperature of small specimens has been developed by the Karlsruhe Institute of Technology (KIT) in Germany. The prototype Creep-Fatigue test module is made by CRPP in Villigen, Switzerland.

Irradiating materials in reactor conditions is a mandatory task to be carried on within the magnetic fusion program which includes ITER and still active third-generation tokamaks. The site for the IFMIF facility is not chosen yet.

6. Controversies

As any project dealing with nuclear energy, ITER faces opponents. From "greens" and environmental non-governmental organisations (NGO) such a protest is just routine. More surprisingly, opposition also aroused within the scientific community. Fusion scientists question the future of tokamaks and the relevance of ITER. A debate began uprising while the initial 1998 ITER project was elaborated. It was made public by the American Physical Society before the design was completed [10]. According to A. M. Sessler (Berkeley) and T. H. Stix (Princeton), too many problems were left unsolved: efficiency of the divertor, incomplete harnessing of the instabilities, damages to the walls, plasma density control... Furthermore, the way innovative concepts could be combined with an otherwise conventional device looked hazardous. On the contrary, M. N. Rosenbluth, a supporter of the project, emphasized the necessity of reactor scale testing.

Nowadays, the scientific community in the US is supportive of the 2001 ITER. The National Research Council issued a positive advice [11] and accordingly the US participated in the funding.

The device under construction at Cadarache is far from the requirements of a power plant. Significant results are expected by the end of the third decade of the 21st century, some 40 years after the project inception. They will still be modest with respect to the objective of a competitive energy source. However according to some scientists, the price tag looks disproportionate. They fear that building ITER at such a high (and still increasing) cost might slow the generation-4 fission programs [12]. To this, it can be replied that the investment is about the same as the amount of money an oil company is paying for implementing a 500000 barrel-per-day oil field.

The stakes are important. The future of magnetic fusion depends upon a successful ITER. Should the program be stopped for any reason, tokamak research would come to an end. Resuming a magnetic fusion program after a few decades would be difficult since skills and know how would be lost. Note that being an international program, ITER is to a certain extent protected.

References

1. W. M. Stacey, *The Quest for a Fusion Energy Reactor: An Insider's Account of the INTOR Workshop*, Oxford (2010).
2. http://www.iter.org/fr/proj/iterhistory. For a genenal account of the project, see http://www.iter.org.
3. J. L. Tuck, *Nuclear Fusion*, **1** (1961) 201.
4. http://www-jt60.naka.jaea.go.jp/english/jt60/project/html/purpose.html.
5. M. Abdou, *Keynote Lecture*, 2nd symposium Global Center of Excellence of Energy Science, Kyoto (2010).
6. E. H. Lee, L. K. Mansur,. *Phil. Mag.* **A61** (1990) 733.
7. E. E. Bloom, S. J. Zinkle, F. W. Wiffen, *Journal of Nuclear Materials*, **329–333** (2004) 12.
8. Z. Oksiuta, M. Lewandowska, K. Kurzydlowski, N. Baluc, *Physica status solidi* (a) **207** (2010) 1128–1131.
9. http://www.frascati.enea.it/ifmif/.
10. *Physics Today*, June 1996.
11. http://fire.pppl.gov/nrc_bpac_draft_prepub.pdf.
12. G. Charpak, J. Treiner, S. Balibar, *Libération*, August 2010 (in French).

7

Some Features of Inertial Confinement — The Role of Lasers

In parallel with tokamak developments, laser-induced inertial confinement proceeded with similar dynamics: rapid progresses during the 1970s and 1980s followed by a plateau which corresponds to the design and the construction of megajoule lasers, instruments that are required for decisive advances.

Inertial confinement fusion (ICF) does exist in nature. Stars, including our Sun are natural fusion reactors. They are made of huge masses of hydrogen compressed due to their own gravitational attraction. As the density increases, so does the temperature until at the centre conditions are met for the onset of thermonuclear reactions. Heat is released. The pressure grows until the pressure gradients exactly compensate gravitational forces. Meanwhile, energy is conveyed via radiative transfer towards the external surface and eventually radiated into the interstellar space. A steady regime is set up for billions of years.

On Earth, gravitational attraction of human size objects is too low by orders of magnitude for a stellar like process. An artificial method is required for compression and a subsequent explosive regime is unavoidable.

The thermonuclear combustion in dense hot matter is to be completed before the medium rapidly blows off. The expansion can be made slower by the inertia of heavy elements driven outwards by the overall motion.

1. Control of a micro-explosion

Controlling a thermonuclear explosion, an exceedingly violent process, is possible only if the reacting mass is limited so that the released energy is below a harmful level. Consequently, micrograms of DT should be compressed up to densities a thousand times the density of liquid matter.

Indeed (see Chapter 2), spherical masses with the same areal density ρr vary as the reciprocal of the density squared. A thousand times the liquid density is comparable with the density at a stellar centre. However, in order to achieve an efficient burn, the temperature should be 10 times higher than that in the Sun. Table 7.1 displays conditions to be met for DT inertial fusion.

Getting the required DT densities is a difficult challenge. Common sense would recommend starting with the highest available density, i.e. solid or liquid state, provided it is made compressible, a requirement to be dealt with in Section 7.2. Now, a $10\,\mu g$ DT liquid sphere is only $160\,\mu m$ in diameter (less than 0.2 mm). This is a very small object. Compressing it adiabatically should be performed in a time hardly longer than the quotient of the radius by the sound velocity in DT at ignition temperature. During these few billionths of a second (usually called *nanoseconds*), a maximum amount of energy is to be given to the fuel from an appropriate driver.

In principle, inertial confinement fusion looks like the way an internal combustion engine works (Figure 7.1): an air–fuel mixture is compressed

Table 7.1. Conditions for a thermonuclear micro-explosion.

Tritium burn fraction f:	> 0.3
Areal density ρr:	$> 20\,\mathrm{kg\cdot m^{-2}}$
DT density:	$> 1000\,\mathrm{g\,cm^{-3}}$
DT mass:	$> 10^{-8}\,\mathrm{kg}\ (10\,\mu g)$
Fusion energy:	$> 1\,\mathrm{MJ}$

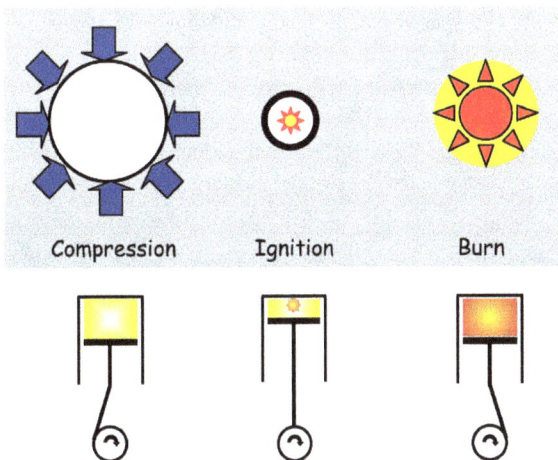

Figure 7.1. Inertial fusion and internal combustion engine. In both cases, the fuel is compressed and after ignition reacts violently. However, the expansion of reaction products produces mechanical work in the cylinder of a motor contrary to the fusion case.

and explodes either after ignition similar to a spark plug or spontaneously as in the case of diesel.

In inertial fusion, the compression is three dimensional. In the motor, it is one dimensional with about the same linear compression rate of 20. Isentropic compression (without any heat exchange with the external word, nor faster than local sound) minimizes the energy invested in the compression process. It is then readily shown that given the final areal density ρr, the required energy to compress a small mass varies as the mass to the 2/3 power (see Box 7.1). On the contrary, the corresponding power is mass independent. For the expected ρr, it is at least 100 TW, a value larger than the installed electric power in many countries. Consequently the driver should be a very high power device.

Box 7.1. Energetics of the isentropic compression

At any time during the compression, the temperature T, the mass density ρ and the pressure p obey the well-known thermodynamic relationships $\frac{p}{\rho^{\gamma}}$ = constant = H, $\frac{T}{\rho^{\gamma-1}}$ = constant, in which the exponent γ depends upon the nature of the fluid. In fusion relevant cases $\gamma = 5/3$. Given the mass to be compressed and the final ρr, the required energy E_C is calculated after these formulas. Let M_0 be the mass, much larger than $10\,\mu g$, of a sphere with the final ρr. Starting at M_0 and decreasing the mass, the compression energy first increases, passes through a maximum and finally tends to zero faster than the mass according to the proportionality $E_C \propto M_0^{2/3}$, a very favourable dependence.

The pressure p is applied to the external surface of sphere with radius R that is moving with the local sound velocity. Accordingly, the power P_M is given by $P_M = 4\pi R^2 p\frac{dR}{dt} = 4\pi R^2 p\sqrt{\frac{T}{\langle m\rangle}} = 4\pi\sqrt{\frac{r_g}{\langle m\rangle}}H^{3/2}(\rho R)^2$ where $\langle m\rangle$ is the average mass of the ions in the fluid, $\gamma = 5/3$ and r_g is the proportionality constant between p and ρT. It turns out that P_M does not depend upon the compressed mass. It is 100 TW for a final ρr of $2\,g/cm^2$.

2. Dynamics of the compression

Getting the required compression rates involves complex dynamical processes. Numerical simulations are necessary for a detailed account of what might happen. They are widely used for the design of experiments: targets and drivers. Codes were devised as soon as researches began on laser interaction with matter [1–3]. However, simplifying assumptions made analytic modelling suited to qualitative investigations [4,5] (see Box 7.2). This leads to a simple formula for the time history of the power to be applied to a sphere:

$$P_M(t) = \frac{P_M(t=0)}{\left(1 - \frac{t^2}{t_F^2}\right)}$$

where t_F is a focusing time. The divergence at t_F is of course non-physical. In real life, available energy and power are bounded. So is the compression. Actually the density in the compressed core reaches a maximum at a time shorter than t_F. Afterwards, expansion begins.

Box 7.2. Homogeneous isentropic compression of a fluid sphere

Assume all fluid elements are compressed according to the same time history. It is then possible to describe analytically the *implosion* of a sphere. Every fluid element at radius r(t) moves towards the centre according to $r(t) = r_0\sqrt{1 - \frac{t^2}{t_F^2}}$ where t_F is a focusing time.

The final complete collapse is non-physical. The model applies only for a duration shorter than t_F. The subsequent motion in which matter bounces back, is to be dealt with by numerical simulations. At maximum compression, part of the kinetic energy is converted into thermal energy: the central temperature rises. Its maximum value increases with the mean implosion velocity.

In liquid or solid state, the DT mixture is hardly compressible. It can be made compressible the following way: a high pressure is suddenly imparted at the outer surface generating a centripetal *shock wave*. A shock wave propagates discontinuities of density, temperature and pressure. Its velocity is supersonic with respect to the motionless medium which is transformed into a compressible dense plasma. The medium is thus adequately prepared for the compression stage.

As shown on Figure 7.2, the instantaneous density profile during the isentropic compression of an initially solid sphere displays a characteristic shape: on upstream of the shock, the density increases with the radius

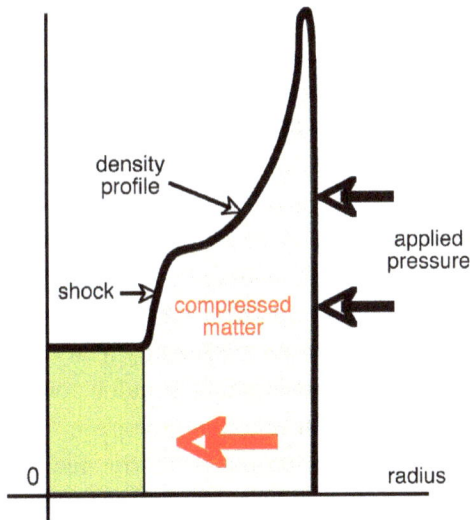

Figure 7.2. Radial density profile during the implosion of an initially solid DT sphere. A high pressure imparted at the outer surface generated a shock wave. Matter is thus made compressible and has acquired a centripetal motion. An adequately tailored pressure time history induces a density peak increasing until all the matter is eventually concentrated at the centre.

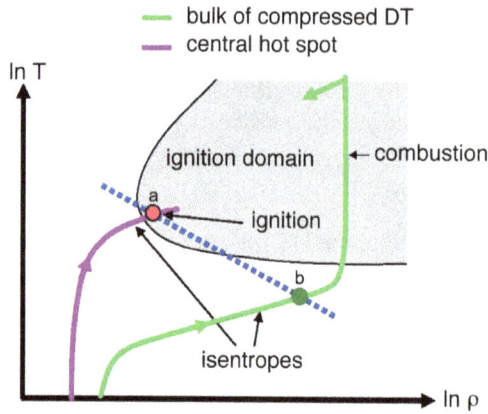

Figure 7.3. Thermodynamic trajectories for central ignition of thermonuclear reactions (after a CEA document): whilst the central zone reaches ignition conditions (point a), peripheral layers follow a low-temperature isentrope up to point b on the same isobaric line as a. A heat wave propagates the thermonuclear burn from the centre towards outer layers. In such a scheme, at least 10% of the fuel react before the sphere is blown off.

according to a parabolic law which also applies to the temperature and to the pressure.

The time history of every layer in the sphere being compressed is to be carefully tailored in order to ignite an explosive thermonuclear reaction in a small mass of fuel. This is best shown graphically. On a temperature density diagram, the behaviour of a given layer is represented by a thermodynamic trajectory, a line in logarithmic coordinates in the case of an isentropic compression (Figure 7.3). For hydrogen isotopes, the slope is 2/3. The final temperature and density conditions to be reached for a spontaneous ignition of the whole mass are represented by a large area in the diagram. Now, if the whole mass were to ignite simultaneously, all layers would have followed the same isentropic. The energy invested in compression would be still too high for a profitable thermonuclear gain. In order to minimize the compression energy, the bulk of the fuel should follow a low temperature isentropic that does not meet ignition conditions at the end of the process. Then the shocked fluid passes through a transient state of quantum degeneracy, a unique situation in fusion plasmas.

When the expected areal density is obtained, ignition does not occur unless it is triggered either externally (the equivalent of the spark plug) or internally at the centre as the detailed compression dynamics make it possible. Indeed the converging shock wave is reflected at the centre with two consequences. First, the reflected shock opposes the centripetal motion of the imploding matter. Second, entropy is introduced locally into the system. Kinetic energy is thus converted into internal energy and the temperature rises significantly, provided the spherical symmetry prevails all along the process. For a rough estimate, ions will have a mean thermal velocity equal to the implosion velocity. In order to get 10 keV at the centre of the compressed DT sphere, the implosion velocity should be 100 km/s.

Numerical simulations have shown that successive converging shocks can be created and give a central *hot spot* with convenient ignition conditions: high temperature and comparatively low density which is an isobaric configuration. There, ignition starts and the thermonuclear burn propagates into the surrounding medium due to a heat wave fast enough to encompass the whole compressed sphere before expansion occurs. Figure 7.3 shows the thermodynamic trajectories followed by the central and the peripheral layers whenever the central hot spot ignition can occur.

An accidental preheat would cause the fuel to follow a higher temperature isentropic. At given invested energy, the compression would be incomplete and the entire process would end in a misfire.

3. Interaction of a laser beam with a solid target

For a successful implosion, a high power should be delivered in a short time. As soon as it was invented, the giant pulse laser appeared as an adequate driver. The performance growth has been impressive from the 1 J, 30 ns tabletop laser of the early 1960s up to the 2 MJ, 0.5 ns versions of the 2010s that are used in attempts to demonstrate ignition.

The laser is a powerful driver with a bonus. The way laser beam interacts with matter is well suited to directly drive an implosion as it was shown in France and elsewhere during the late 1960s [6].

When a sufficiently intense laser beam is focused onto a solid surface, a plasma plume, also dubbed corona, is created and flows outwards into the surrounding vacuum. A rarefaction wave is formed. The radiation is absorbed in the density gradient. At optical frequencies, the critical electron density is well below the electron density inside a metal. The wavelength of a neodymium glass laser is $\lambda = 1\,\mu m$, its circular frequency is $\omega = 1.8 \times 10^{15}\,s^{-1}$. The critical or cut-off electron density is $n_c = 10^{21}\,cm^{-3}$, well below $5 \times 10^{22}\,cm^{-3}$ in a metal. In the plasma flow, the cut-off separates an *underdense zone* through which the laser beam propagates from a forbidden *overdense zone* close to the solid. Inside the overdense zone, energy is transported by heat conduction, hence the alternative name: *conduction zone*. However, energy does not propagate beyond an ablation front at which the temperature is low whilst the temperature gradient tends to infinity. The ablation front acts also as a *thermal barrier*.

Performing a quick energy balance calculation, it is readily shown that the pressure in the overdense zone is very high: at least several megabars. This is orders of magnitude larger than in the unperturbed medium. A shock wave is thus a necessary intermediate. Hence the general scheme sketched in Figure 7.4 of the plasma flow induced by a laser beam irradiating a solid surface. Now, momentum should be conserved. Consequently the dense part should move in the direction opposite to the outwards plasma flow in the corona.

Therefore, a kind of rocket effect imparts momentum to the dense compressible plasma squeezed between the shock wave and the ablation front. The subsequent motion is obviously favourable to the compression

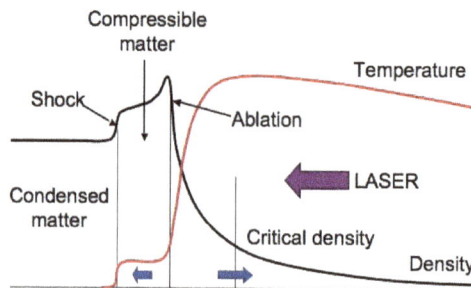

Figure 7.4. Temperature (in red) and density (in black) profiles induced by a high-intensity laser beam impinging onto a solid surface. The coronal plasma is to the right, starting at the ablation front. Laser radiation does not propagate beyond the critical density. Matter squeezed between the shock wave and the ablation front moves towards the left hand side whereas matter carried out in the rarefaction wave moves towards the right hand side (blue thick arrows).

Chapter 7. Some Features of Inertial Confinement — The Role of Lasers

Table 7.2.	Interaction regimes according to the values of the interaction parameter $I\lambda^2$.		
Regime	$I\lambda^2$	T	Applies to
Low intensity	$I\lambda^2 < 10^{12}\,\mathrm{Wm^2 cm^{-2}}$	$T \propto (I\lambda^2)^{1/3}$	Industry
Linear	$10^{12} < I\lambda^2 < 3 \times 10^{13}\,\mathrm{Wm^2 cm^{-2}}$	$T \propto (I\lambda^2)^{2/3}$	Fusion: compression
High intensity	$3 \times 10^{13}\,\mathrm{Wm^2 cm^{-2}} < I\lambda^2$	$T \propto (I\lambda^2)^{1/3}$	Fusion: fast ignition

of thermonuclear fuel according to the dynamics described in Section 7.2. Altogether, the density profile looks very similar to the density profile in Figure 7.2.

Physicists need numbers in order to deal with observations and calculations. In the case of laser interaction with bulk matter, an *interaction parameter* was defined: the product of the intensity at the target surface times the wavelength squared, i.e. $I\lambda^2$. A plasma is created provided the interaction parameter exceeds some $10^7\,\mathrm{W\mu^2 cm^{-2}}$ (usual practical units). Collecting the results of many experiments, three interaction regimes are shown according to the values of the interaction parameter in Table 7.2.

Below $10^{12}\,\mathrm{W\mu^2 cm^{-2}}$, volume absorption of the laser light occurs inside the underdense zone. The plasma temperature varies as the 1/3 power of the interaction parameter. The pressure is too low for a compression of interest for fusion. From 10^{12} to $3 \times 10^{13}\,\mathrm{W\mu^2 cm^{-2}}$, the regime is linear. Light is absorbed at cut-off and the temperature is found to vary as the 2/3 power of the interaction parameter. Above $3 \times 10^{13}\,\mathrm{W\mu^2 cm^{-2}}$, nonlinear effects appear (see Section 7.4 for details). A large part of the absorbed energy goes to plasma oscillations instead of heating the plasma. This unfavourable effect precludes high pressures. However this regime might be useful for fusion since it might contribute to the fast ignition concept that will be dealt with in Chapter 8.

In the very special conditions of the linear regime, light absorption takes place in a thin layer at cut-off. At this point, the flow is sonic. Important quantities can be calculated analytically under simple assumptions. The structure of the conduction zone turns out to be stationary. Solving the heat equation for a plasma, a temperature profile is found that is representative of a nonlinear heat transport coefficient (see Chapter 3). The pressure at the ablation front is about twice its value at cut-off.

Although it applies only to a restricted interval, an interaction parameter between 10^{12} and a few $10^{13}\,\text{W}\mu^2\text{cm}^{-2}$, the linear regime is well suited for inertial fusion. The laser energy is optimally used in plasma heating. Either the pressure is high enough to drive a spherical target compression (direct drive) or the secondary thermal radiation emitted by the corona can be used for an indirect drive (see Section 7.5). Pressure and/or secondary radiation fluxes are readily calculated. They are satisfactorily mastered in experiments. The wavelength of the impinging radiation should be chosen in order to remain in this regime as long as possible.

It is possible to increase the impinging laser intensity following a prescribed time history. Consequently, the applied pressure at the ablation front rises and so does the density of the shocked medium close to the ablation front. An almost isentropic compression is thus obtained due to the thermal barrier effect. About 10% of the laser energy goes into the kinetic energy of the shocked and compressed matter. In spherical geometry, matter moves at high velocity towards the centre. The flow obtained in laser interaction with the entire surface of a solid sphere matches the requirement for inertial confinement fusion.

4. Instabilities

Laser-induced inertial fusion is not free of instabilities. However, their nature is different from instabilities encountered in magnetic confinement. They can be classified into two categories: instabilities in laser plasma interaction, instabilities in fluid motion.

Whenever the interaction parameter exceeds a threshold of order 2 to $3 \times 10^{13}\,\text{W}\mu^2\text{cm}^{-2}$, nonlinear couplings between the incoming laser waves and plasma oscillations occur. Such phenomena are known as *parametric instabilities*. They are enhanced by resonance effects when the laser frequency is close to the electron plasma frequency (at cut-off) or close to twice this value.

Once instabilities are fully developed, the laser energy is no longer transferred to plasma thermal energy and pressure. Instead energy goes to oscillation modes and superthermal electrons. Furthermore laser light is scattered by plasma waves: *Raman scattering* on electron plasma waves;

Brillouin scattering on ion-acoustic waves. In both cases, backscattering is the worst possible mechanism: a large part of the incoming energy is sent back towards its source. Other effects stem from the coherence of the laser light. The energy density of an electromagnetic wave acts on matter via a ponderomotive force. Since the plasma is a fluid, modulations of the plasma density build up and tend to be spontaneously organized as a diffraction grating consistent with the laser wavelength. In extreme cases, the structure grows unstable and filamentation eventually sets chaos in the interaction zone.

Hydrodynamic instabilities deal with the motion of interfaces. They induce deviations from the spherical symmetry with two consequences: preventing a high compression of the fuel and/or mixing it with heavier materials, thus enhancing radiation emissions. As a result, the reaction rate would be less than expected rendering ignition all but impossible.

First, consider two liquids in a gravitational field, the heavier one above the lighter one. Given any perturbation of the interface, heavy fluid fingers develop downwards across the lighter fluid and eventually turbulent motions lead to mixing of the two fluids. This is the well-known Rayleigh–Taylor instability. In laser driven inertial fusion, it occurs in regions where inertial forces are equivalent to a gravitational field. This is the case in the rarefaction wave when the ablation rate is increasing. Another case deals with the DT enclosed inside a glass shell. At the end of the implosion process, the heavy shell is decelerated while pushing the compressed central light fuel.

Another instability arises when a shock wave crosses an interface. The initial distortions are to be amplified. This behaviour is known as the *Richtmyer–Meshkov* instability.

Hydrodynamic instabilities plague the implosion process. A detailed knowledge of their dynamics is needed to overcome them. To this end, analytical theory is not powerful enough. It is completed by state-of-the-art numerical simulations and dedicated experiments.

In order to study these instabilities, specially designed targets are irradiated by adequately tailored laser pulses. The main diagnostic in such experiments is X-ray instant radiography, the source for backlighting being a laser irradiated plate. The necessary magnification is secured by grazing

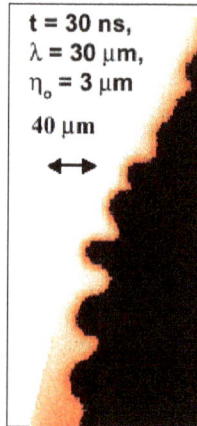

Figure 7.5. After P. Munsch's doctoral dissertation [7]. Instant radiography of a planar double-layered target, Cu (in black)-CH_2 (in orange — false colour), 30 ns after a laser-pulse impact (Phébus, see Table 8.1). The interface was initially corrugated. The wavelength λ of the defect was 30 μm and its amplitude η_0 was 3 μm grown up to 30 μm after 30 ns.

incidence X-ray optics. Figure 7.5 is an example of results obtained in a typical experiment.

X-ray backlighting does not provide reliable information about the interior of compressed spherical objects. In order to get some insight, experiments on Rayleigh–Taylor instabilities were performed in cylindrical geometry at the Nova laser facility (see Table 8.1). A series of frames displayed in Figure 7.6 shows the implosion of a cylindrical shell with an initial 12 fold azimuthal defect.

5. Indirect drive

Neodymium glass lasers are the most powerful to date. They radiate in the near infrared at wavelength $\lambda = 1.06\,\mu$m. Even after tripling or quadrupling, the frequency is still too low for a linear interaction at the intensities that correspond to the areal density ρr the implosion process is aiming at.

The challenge of inertial fusion requires an implosion as spherical as possible, specially at the end of the process in order to guarantee central ignition. Now, with laser beams focused at the sphere outer surface, the

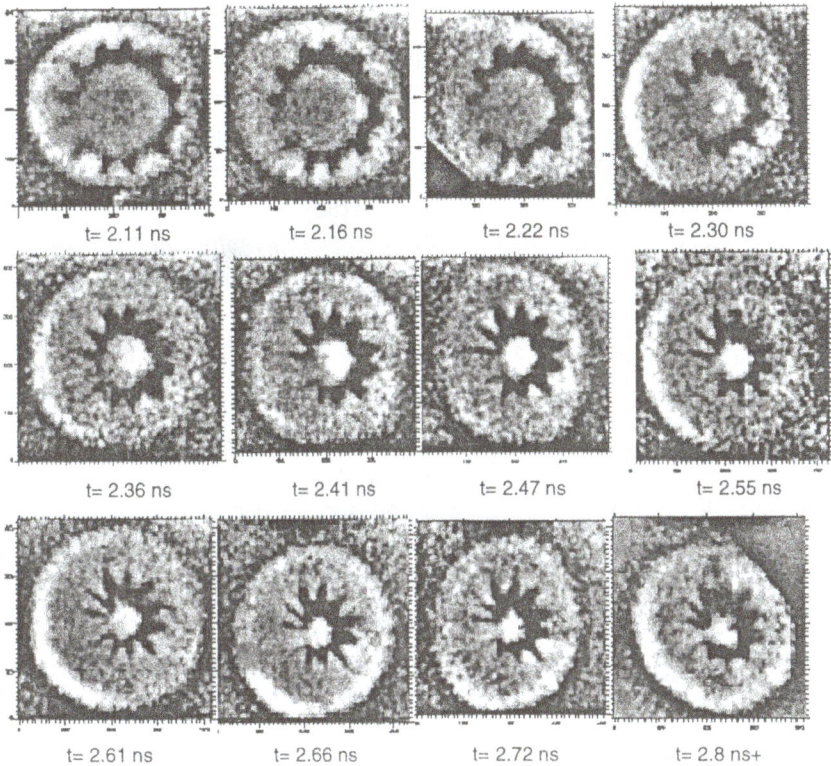

Figure 7.6. Rayleigh–Taylor instabilities in an imploding cylindrical target [8]. Polystyrene foam that does not interact with backlighting X-rays is enclosed inside a double layer cylindrical shell. The inner layer is an opaque marker which appears in black in the frames. The external surface of the outer layer has a dodecagonal cross section thus providing the initial defect. During the radiatively driven implosion, the ablative instability develops and fingers are seen penetrating the plasma rarefaction wave. The slowing down of the shell induces another Rayleigh–Taylor instability with fingers penetrating the compressed foam.

illumination is far from uniform, even in the case of megajoule lasers with about 200 beams. Due to hydrodynamic instabilities, illumination defects are amplified during the implosion. The spherical symmetry is destroyed which prevents high compression and subsequent important thermonuclear burn.

Such drawbacks can be avoided with indirect drive. Indeed, the plasma plume in the interaction of a laser beam with a metallic surface is a strong X-ray emitter with a near blackbody spectrum. The trick is the following: the sphere to compressed is at the centre of a cylindrical cavity (dubbed "*hohlraum*") with open ends through which laser beams enter and interact with the internal surface. The secondary X-rays from the plumes invade the cavity inside which they are trapped. Materials and laser intensity are chosen so that the radiation temperature is a few hundreds of eV. The thermal radiation has a uniform intensity at the external surface of the central capsule coated with an ablator (see next section) with which interaction occurs and an implosion is initiated. The interaction scheme exhibits the same density and temperature profiles as presented in Figure 7.4. The only difference is the absence of a critical density.

The comparison between direct and indirect drives is given in Figure 7.7. The intensities needed for an efficient X-ray emission from the plasma plumes are far below the requirements for directly driving the ultimate step of an implosion. Parametric and backscattering nonlinear effects are avoided to a certain extent. The cavity is usually cylindrical. Typical figures for its overall length and diameter are 5 to 10 mm and 3 to 5 mm respectively.

As far as the illumination uniformity of the central sphere is concerned, the indirect drive has proved successful. The final volume is fairly spherical. Unfortunately, in order to achieve this result, the efficiency of the entire compression process is lowered since only 30 to 50% of the incoming laser energy is converted into thermal radiation.

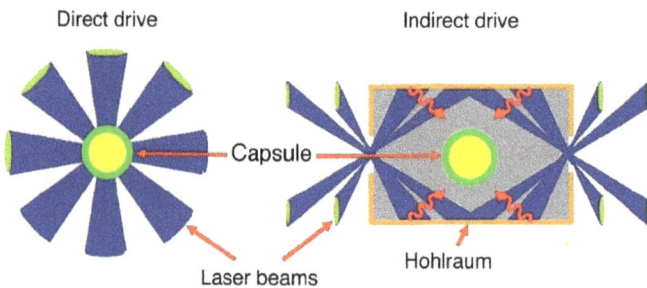

Figure 7.7. (CEA document). Laser beam geometry: (a) in direct drive, they converge in order to impinge onto the external surface of a spherical capsule; (b) in indirect drive, they interact with the inner surface of a cylindrical cavity containing the spherical capsule at its centre. The secondary radiation from these interactions (wavy arrows) is to perform the compression of the thermonuclear fuel.

6. Target design

Early implosion experiments were performed using a target glass *microballoons* filled with gaseous DT at pressures a few tens of bars. Irradiated by intense laser pulses, the thin glass shell explodes. Part of it expands outwards violently. The remaining part expands inward thus compressing and heating the inner gas.

Optimizing the implosion requires a more sophisticated design (Figure 7.8). The outer shell is a hydrogen-rich plastic *ablator* to be transformed into a plasma flowing outwards after interaction either with laser light (direct drive) or secondary thermal radiation (indirect drive). Then a thin shell is made of a metal with high atomic number Z (gold is widely used in state-of-the-art experiments). It is to serve as a thermal barrier. It separates the core to be compressed from the outer plasma corona thus preventing fuel preheating by fast electrons produced by nonlinear radiation plasma interactions. In the case of a cryogenic capsule, the inner core is a glass microballoon containing a thick layer of liquid DT whose compression up to a very high density is expected. The central DT bubble is to be shock heated at the end of the implosion process and serve as an *ignitor* for the thermonuclear burn. Matching each other, the overall diameter, the thicknesses of the different layers, the implosion dynamics and energetics are determined thanks to detailed numerical simulations.

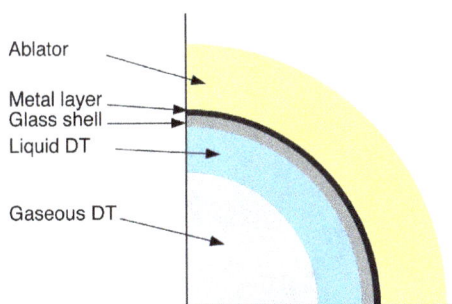

Figure 7.8. Target design for a direct drive implosion. Each concentric layer is dedicated to a specialized function. The thin metallic layer coating the glass microballoon enhances the thermal barrier effect between the hot dense shocked ablator and the DT to be compressed isentropically. The central bubble is to serve as an ignitor. The glass in the microballon is often replaced by a plastic shell.

The glass microballoon is the basic structural element of the target. Tiny hollow sphere a few tenths of a millimetre in diameter and a few hundredths of a millimetre thick are a standard industrial product. They are fabricated the same way as soap bubbles by blowing air trough molten glass. After cooling a small proportion meets the requirements for implosion. By automated processes, a few correct spheres among a million are sorted out: as spherical as possible, air tight and with the right diameter and thickness. DT filling is obtained by high-temperature gaseous diffusion through the glass shell.

Partial gas condensation on the internal surface of the shell is obtained by progressive cooling down to liquefaction temperature (Figure 7.9). Natural properties of matter such as liquid vapour equilibrium, surface tension, radioactivity (see Box 7.3) are effective in the formation of a smooth uniform layer.

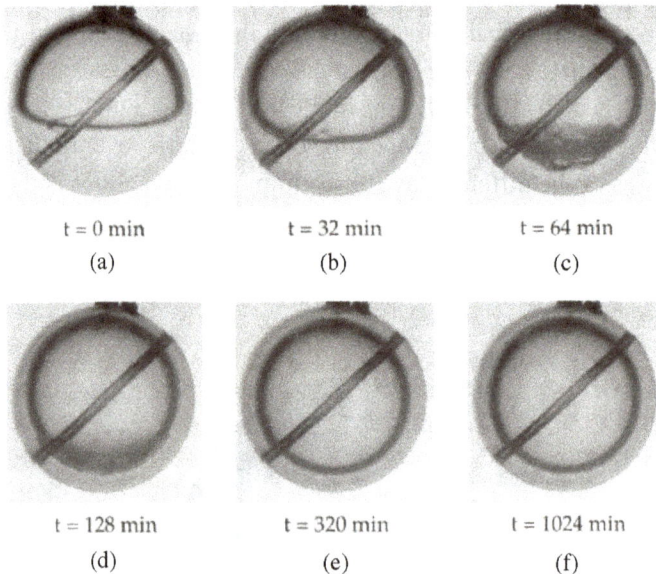

| t = 0 min | t = 32 min | t = 64 min |
| (a) | (b) | (c) |

| t = 128 min | t = 320 min | t = 1024 min |
| (d) | (e) | (f) |

Figure 7.9. (LLNL document). DT condensation inside a glass microballoon filled with gas. Cooling is progressively driven in a cryostat at liquid hydrogen temperature. Condensation begins in the lower hemisphere. Then as the liquid invades the cavity, a meniscus is formed. Eventually the meniscus is transformed into a bubble automatically centred by surface tension forces.

Centering of the inner bubble in a microballoon is made stable by the β radioactivity of tritium. Since their mean free path is very small inside the liquid, emitted electrons induce local heating. Should a bump form at the surface, heating and evaporation are enhanced. Since the cryostat is steadily at 19 K, the excess vapour condensates uniformly onto the internal liquid surface.

Since they include a hohlraum, targets for indirect drive are complicated, but the design of the central capsule can be made simpler. Preheating of the DT by fast electrons is less likely than in direct drive. Consequently, a metallic thermal barrier is no longer necessary. Furthermore it is nowadays possible to manufacture plastic microballoons with convenient mechanical properties.

More involved is the hohlraum cavity. Thermal radiation and plasma should be confined with a high pressure whose time history is to be adjusted to the implosion dynamics. The container is made of metal or glass coated with a metallic layer. The capsule is held by a glass stalk. Laser beams enter through the axial apertures and impinge onto the inner surface whose material is chosen in order to ensure a large emission of secondary radiation (Figure 7.10).

For implosion experiments to be performed with the latest generation of high-power lasers, the targets design will incorporate cryogenics and

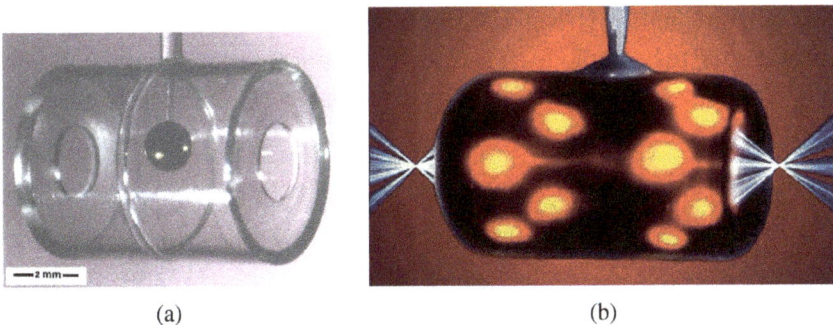

(a)   (b)

Figure 7.10. Targets for indirect implosion: (a) a microballoon inside a glass-made cylindrical cavity (CEA document). (b) Thermographical visualisation (false colours) of laser beam impacts on the internal surface of the cavity generating heat waves that have propagated across the cavity wall (LLNL document).

control devices aimed at keeping the temperature of the solid or liquid DT layer at the right value during transportation and setting inside the target chamber. High-level science and know how are obviously required for such achievements.

7. Compression experiments

Laser-driven inertial fusion relies on pulses whose time history is tailored in such a way that the pressure applied to the capsule grows as required for an efficient compression of a DT mixture up to more than 1000 times the liquid density. Early compression experiments were performed with available lasers and simple targets according to the direct drive scheme. As time elapsed, more powerful multibeam laser pulses were used together with advanced design targets. After the completion of the National Ignition Facility (NIF) at Livermore in 2009, an unconclusive first attempt at ignition (National Ignition Campaign) was conducted in 2011–2012.

In compression experiments, the size of the central sphere is to be monitored. As for the visualisation of hydrodynamic instabilities, the usual diagnostic is X-ray backlighting. An auxiliary laser beam irradiates a metallic plate (titanium is a good choice) providing the necessary X-ray source. In the case of indirect drive, the cylindrical wall is fitted with apertures whose diameter is slightly larger than the capsule diameter. The capsule is thus made visible at the expense of the efficiency of the indirect drive. The contrast is enhanced by seeding germanium in the spherical plastic shell to be imploded. Pinhole cameras are used for imaging. Frames are recorded using an ultrafast camera (Figure 7.11).

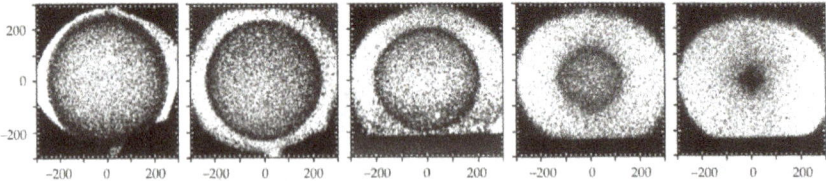

Figure 7.11. X-ray radiographs (backlighting) of an imploding indirectly driven microballoon with initial diameter 400 μm (LLNL document). In order to obtain this sequence, the exposure time was 55 ps. Two successive frames are separated by 0.5 ns.

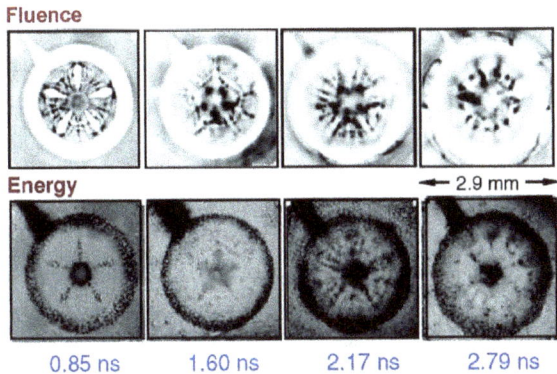

Figure 7.12. Proton radiographs of an implosion [9]. 14 MeV protons from the $D-{}^3$ He reaction travel through the cavity along directions parallel to the axis. There is no need for extra apertures in the cylindrical wall. On each frame the local fluence and the total energy loss of the protons along their trajectories can be measured. The darkening increases as the fluence or the energy loss are larger. The star shaped pattern is a consequence of the disposal of the laser impacts onto the inner surface of the cavity in this particular experiment. The geometry of the plasma flows is thus imposed.

Recently, proton radiography was used as a diagnostic (Figure 7.12). The protons for backlighting are produced by the deuterium–Helium3 nuclear reaction inside a plasma created by direct drive of a spherical target under the impact of auxiliary laser beams [9]. As a supplement to implosion dynamics, this technique provides informations about the plasma inside the hohlraum cavity.

Informations about the plasma inside the compressed core are derived from measurements of the number of neutrons released by the D–T reaction and their spectral distribution. In the case of an indirect drive, the number depends upon the radiation temperature inside the hohlraum cavity. According to numerical simulations, the cavity temperature shows a maximum 1 ns after the beginning of laser interaction (Figure 7.13).

The maximum temperature in the cavity is the relevant quantity for parametric studies associating experiments and numerical simulations. In the experimental results of Figure 7.14, the fuel in the core was pure deuterium. The number of neutrons from the D–D reaction and the ion temperature derived from the neutron spectra are shown in the figure to agree fairly well with calculations.

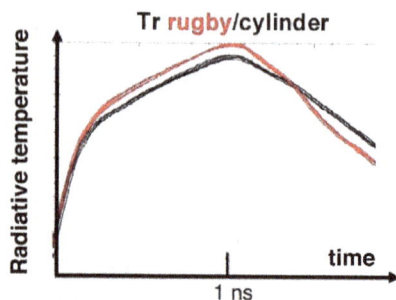

Figure 7.13. (after a CEA document). Numerical simulation of the radiation temperature (Tr) time history for two different hohlraums: in black cylindrical cavity; in red cavity with a rugby ball shape. The maximum temperature is of order 250 eV.

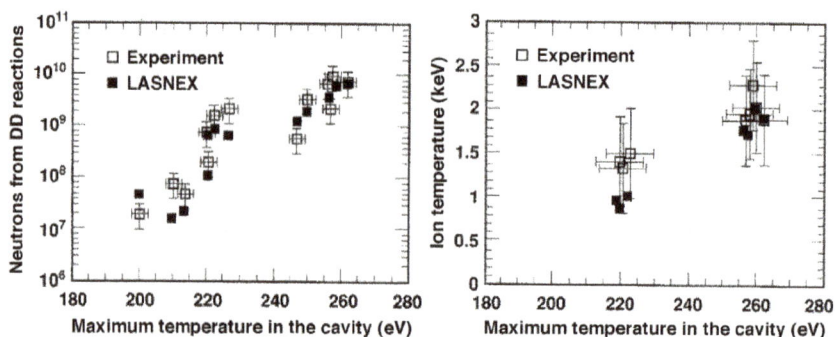

Figure 7.14. Experimental results compared to simulations (after a LLNL document). Pure deuterium in microballoons was compressed by indirect drive. The results from runs of the LASNEX code are found to be close to the experimental data.

Progress towards high densities is visible on the plot of Figure 7.15: densities obtained at the Institute for Laser Engineering (ILE) in Osaka versus calendar time. The first indirectly driven implosions were performed in 1982 using the laser Gekko XII, a large device well-suited to such investigations.

A thousand times the liquid density $(200 \, g/cm^3)$ is a symbolic benchmark. The performance was achieved in 1994 using a cryogenic target. This is at the lower boundary of the domain sought for micro-explosions of thermonuclear fuel. In further experiments, evidence was obtained of even higher densities exceeding $1000 \, g/cm^3$.

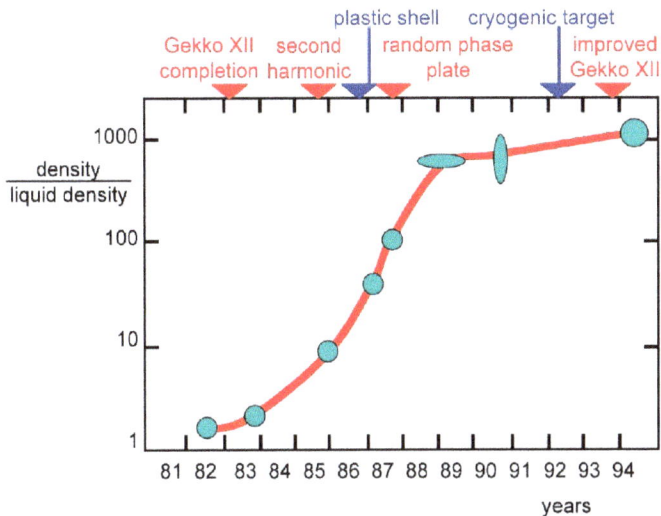

Figure 7.15. (after an ILE Osaka document). Results follow the improvements of the laser (in red) and advances in target design (in blue). With an upgraded GEKKO XII, a density a thousand times greater than liquid state was achieved in 1994.

References

1. J. H. Nuckolls, L. Wood, A. Thiessen and G. Zimmerman, *Nature* **239** (1972) 139.
2. K. A. Brueckner and S. Jorna, *Rev. Mod. Phys.* **46** (1974) 325.
3. J. L. Emmett, J. Nuckolls and L. Wood, *Scient. Am.* (1973).
4. R. E. Kidder, *Nuclear Fusion* **14** (1974) 53.
5. J. L. Bobin and J. M. Reisse, *Rev. Phys. Appl.* **11** (1976) 497.
6. F. Floux *et al.*, *Phys. Rev. A* **1** (1970) 821.
7. P. Munsch, doctoral dissertation, University Pierre et Marie Curie, Paris (2000).
8. Reproduced in B. L. Alcock, MS dissertation, Purdue University (1996).
9. C. K. Li *et al.*, *Science* **327** (2010) 1231.

Big Drivers for Inertial Fusion

Directly or indirectly driven, laser-induced inertial confinement fusion relies on one and the same sequence of events: compression until ignition then thermonuclear burn (Figure 8.1). Since the 1970s, the question has been: what are, in either case, the required laser performances?

Advances in experiments come slowly. Meanwhile more and more sophisticated numerical experiments were performed. Numerous parametric studies were carried on in order to optimize the process. Troublesome mechanisms were accounted for mainly the instabilities already dealt with in Chapter 7: parametric and scattering instabilities in laser plasma interaction, hydrodynamic instabilities of interfacial motion. They have opposite effects (Figure 8.2). It turns out that cumulated effects are minimized for laser energy and power exceeding 1 MJ and 400 TW, respectively.

According to numerical simulations, a transition is expected for a $0.35\,\mu$m laser energy slightly below 1 MJ, as shown on Figure 8.3 which refers to cryogenic targets with a central gas bubble. The *gain*, ratio of fusion energy to the energy of the laser beams is plotted as a function of the latter. In the vicinity of 1 MJ, the gain increases abruptly by at least two

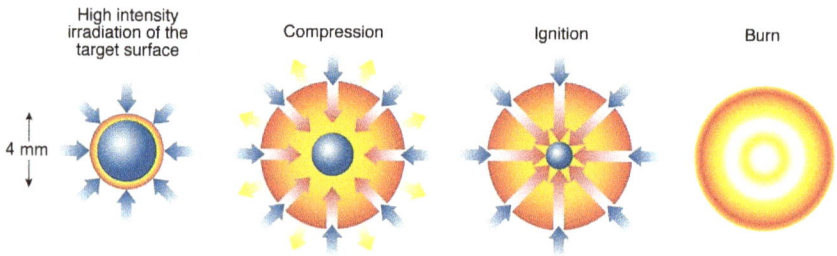

High intensity irradiation of the target surface

4 mm

Compression

Ignition

Burn

Figure 8.1. Steps of inertial confinement fusion. Under the pressure induced by a high radiation intensity (blue arrows), the capsule containing DT is compressed until ignition of the thermonuclear reaction occurs. To be effective the process requires a spherical symmetry as perfect as possible during the compression phase. Red arrows represent the thermal flux. Yellow arrows represent the coronal plasma motion.

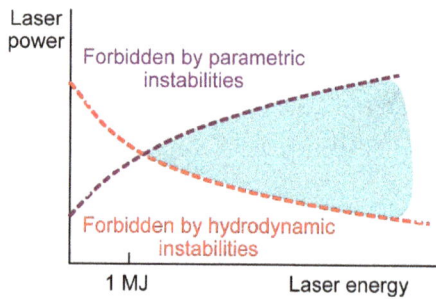

Laser power

Forbidden by parametric instabilities

Forbidden by hydrodynamic instabilities

1 MJ   Laser energy

Figure 8.2. On this power/energy diagram, the coloured area is the domain in which troubles due to instabilities are minimized. At given energy, the power is to stay below an upper limit in order to be free of parametric instabilities. Conversely, rising simultaneously energy and power increases the safety with respect to hydrodynamic instabilities.

orders of magnitude. The transition energy is lower when the compression velocity and the *in-flight aspect ratio* (radius divided by the thickness of the shocked dense layer) are increased.

The power of the laser beams is the crucial parameter. The average implosion velocity, hence the temperature of the central gas sphere are determined by its value. From this hot spot, a reactive heat wave propagates in the surrounding compressed medium whose areal density ρr, also power dependent, should exceed 2 g/cm^2. A minimum energy is also required for an optimized compression. The question is then, what are the laser power and energy that would guarantee ignition?

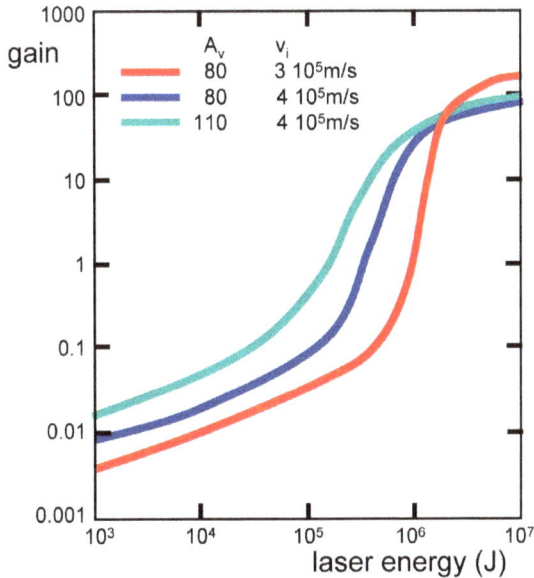

Figure 8.3. (ILE Osaka document). Gain of indirectly driven DT capsules, versus laser energy for several implosion velocities v_i and in-flight aspect ratios A_v. Such numerical simulations evidence a drastic change in the vicinity of 1 MJ.

In order to test the indirect drive, scientists from the Lawrence Livermore National Laboratory and the Los Alamos National Laboratory (both institutions active in the military side of nuclear energy) used a nuclear explosion as a thermal radiation source. In the early 80s, small DT capsules were irradiated during underground tests. According to the scarce pieces of information released in 1988, years after such experiments (dubbed "Centurion/ Halite") were performed, 10 (lower limit in such conditions) to 100 MJ of radiation deposited inside a hohlraum do trigger the thermonuclear burn of DT fuel.

After decades of intensive research, scientists have accumulated a fair amount of knowledge about the physics of both domains: laser plasma interaction and target compression. The success of inertial confinement fusion relies on implementing drivers with the required power and energy. While lasers can be used both for direct or indirect drive, other devices also prove interesting for the indirect drive: heavy ion accelerators and pulsed power generators to be dealt with at the end of this chapter.

2. High-power lasers

Lasers are remarkable light sources. Stimulated emission occurs along a well defined direction. Furthermore, as light is amplified, the width of the emitted line decreases, while temporal and spatial coherence increase. Consequently, a laser beam is close to the ideal picture of a plane monochromatic wave train. Since the days of Archimedes' burning mirrors, mankind knows how to concentrate a light flux. The more directional the light, the better is the focusing. In the case of lasers, focusing produces exceedingly high light intensities.

Box 8.1. The laser effect

In the transition an electron undergoes between two energy levels inside an atom or a molecule, light can be emitted either spontaneously or in a stimulated way in the presence of light with the same wavelength (a process postulated by Einstein in 1916). Stimulated emission opens the way to light amplification, hence the acronym *light amplification by stimulated emission of radiation*. A necessary condition is the overpopulation of the upper level of the transition. The *population inversion* is obtained by *pumping*, through a lot of possible processes. Optical pumping is the most widely used. The active medium absorbs radiation either from an extended spectrum or tuned to a specific transition.

An o*scillator* is made of an active medium inside a Perot–Fabry cavity tuned to the transition to be stimulated (optical resonator). Outside the cavity the active medium usually serves as a single pass amplifier. Multiple pass amplification was recently implemented on the most powerful lasers.

The 50th anniversary of the first laser emission was celebrated in 2010. This landmark experiment was performed by Theodore Maiman irradiating with a helical flash lamp a ruby cylinder with silvered ends.

The first experiments on laser interaction with plasmas used the so-called "giant pulse" lasers (0.1 J, 30 ns) that had been made available in the early 60s: ruby followed by neodymium glass were the active materials. In the case of ruby, it has been so far impossible to grow large single crystals at a reasonable cost, hence a severe limitation of the energy and the power to be extracted from a ruby laser. On the contrary large pieces of glass with good optical qualities are commonly produced. The neodymium glass technology can be extrapolated up to the energy and power required for significant fusion experiments to be performed in the 2010s. However the

Table 8.1.	45 years of Neodymium glass lasers.				
Year			Energy	Pulse duration	Number of beams
1968	L5	Limeil (CEA France)	100 J	5 ns	2
1973	C6	Limeil (CEA France)	300 J	3 ns	6
1977	Shiva	Livermore (USA)	10 kJ	0.5 to 1 ns	12
1980	Octal	Limeil (CEA France)	1 kJ	1 ns	8
1980	Omega	Rochester (USA)	2 kJ	ns	24
1983	Gekko XII	Osaka (Japan)	10 kJ	1 to 2 ns	12
1985	Nova	Livermore (USA)	60 kJ	3 ns	10
1985	Phébus	Limeil (CEA France)	6 kJ	1 ns	2
1995	Omega	Rochester (USA)	30 kJ	ns	60
2009	NIF	Livermore (USA)	1.1 MJ*	ns	192
2014?	LMJ	CESTA (CEA France)	MJ	ns	240

*After frequency tripling. Result obtained in March 2009.

wavelength (1 μm) is incompatible with a linear interaction regime at the required intensities. To avoid such a setback, frequency doubling, tripling or quadrupling are routinely implemented.

Starting from less than one joule and megawatts, it took fifty years of technological improvements to achieve the megajoules and terawatts available to date. The race towards gigantic lasers (Table 8.1) began in the early 60s, once it was recognized that temperature of thermonuclear interest were to be obtained by focusing laser light on suited targets.

The basic design was proposed in the 1960s. As shown on Figure 8.4, a laser system aimed at fusion is made of three parts:

- A compact *front end* generates low-level pulses: oscillator and pulse shaping devices combined with preamplification stages.
- Parallel amplification of a single pulse up to high power by splitting and amplifying each branch. The beam diameter increases from one stage to the following one so that the active medium does not undergo excessive thermal and radiative stresses. The beam is filtered between the stages in order to keep a good uniformity all along the optical path, a few hundred meters.
- Optical elements for beam transport, frequency conversion and final focusing.

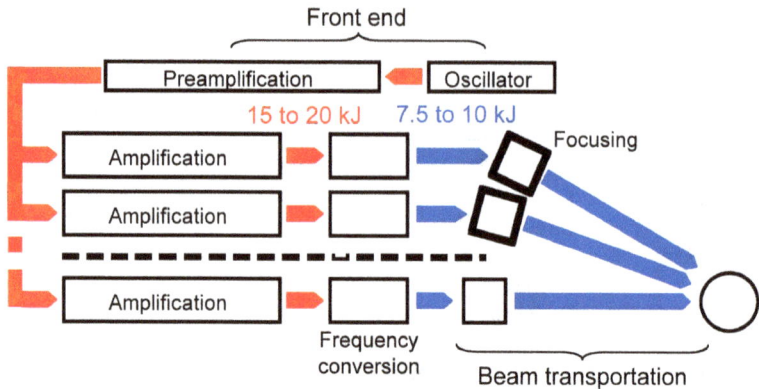

Front end

Preamplification ← Oscillator

15 to 20 kJ   7.5 to 10 kJ

Amplification → [ ]   Focusing

Amplification → [ ]

Amplification → [ ] → [ ]   ◯

Frequency
conversion   Beam transportation

Figure 8.4. Geometry of a high-power laser for fusion. The front end delivers a 1 J pulse. Then, the beam is divided before being amplified by many chains in parallel. Energies stated before (in red) and after (in blue) frequency conversion refer to a single high-power chain.

The configuration stems from amplification dynamics. At low intensity, the signal travelling through active material grows exponentially with respect to the thickness. It is amplified by orders of magnitude. At high intensity, amplification is at best linear. In a laser chain, functions are implemented separately: pulse creation and tailoring at low level, linear amplification close to saturation in high-power parallel chains (Figure 8.5).

Intense repetitive laser fluxes induce damages within materials. Many effects contribute to the damages whenever the energy flow per unit surface (fluence) is large. The refractive index of a transparent medium is intensity independent only if the flux density is low: linear regime in which the polarizability is proportional to the electric field of the light. At high intensities the polarizability is no longer linear. Self focusing generates locally exceedingly high intensities the material cannot withstand without damages. Furthermore, in active materials filamentation occurs causing a spatial disorganisation of the laser beam.

Impurities are another reason for a poor strength. Light absorbing local inhomogeneities cause thermal stresses. Worse when fusing neodymium glass, metal inclusions might migrate from the crucible. This setback happened with platinum crucibles, a metal suited to obtaining homogenous glass but metal inclusions could not be avoided.

A wavelength dependant damage threshold exists for every transparent material. It is usually determined experimentally. Along an amplifying

Figure 8.5. (CEA document). High power chains being installed for the Laser Mégajoule (LMJ). Between optical elements the laser beams are travelling inside cylindrical pipelines in order to avoid atmospheric convection and turbulence effects.

chain the beam diameter increases in order to keep the flux density well below threshold. Half a meter is a convenient upper value of the transverse size of manageable optical elements. Consequently the laser energy that can be extracted from a single high-power chain is at most 20 kJ. Building a megajoule instrument requires a large number of chains in parallel.

In the front end, active materials doped with neodymium are first single crystals of yttrium aluminium garnet (YAG) then glass rods. Active elements along the high-power chains are made exclusively of glass. Altogether, the total thickness of active materials is a small fraction of the distance from the oscillator to the target. Indeed the gain depends upon the laser beam quality which is improved by filtering devices incorporating lengthy paths.

Optical pumping is obtained using flash lamps. A small fraction of their spectrum corresponds to the absorption bands of neodymium imbedded in garnets or glass. Although the process is inefficient, it is widely used since industry is able to provide reliable equipment. No other technology so far has proven an equal capability at the power level required for inertial fusion.

In the neodymium glass, the absorption length for the radiation emitted by flash lamps is at the order of 5 cm. Most of the radiative energy is converted locally into heat. Now the glass needs time for cooling, typically tens of minutes up to hours for the biggest elements. A long time interval is thus required between successive shots. For the most powerful lasers, one or two shots per day is maximum. The rather small absorption length has consequences on the design of laser amplifiers. The diameter of a glass rod pumped sideways cannot exceed 10 cm. The trick for larger diameters was invented circa 1970. Slabs or disks (Figure 8.6) are used instead of rods. Cross-sectional beam diameters up to 80 cm were achieved thanks to this technology.

The disks or plates used are at most 5 cm thick. The laser beam is incident at Brewster angle on antiparallel pairs of plates. Light from the flash lamps penetrates the plates through the plane surfaces in order to ensure a homogeneous pumping. Satisfactory propagation and amplification of the laser beam require a high optical quality of the plates. 80 cm in diameter as in the "Nova" laser is a maximum size. In newly built megajoule lasers, each beam has a diameter slightly below 40 cm and carries at most 15 to 20 kJ.

In a high-power laser, pulses are delivered by an oscillator at a constant repetition rate, e.g. 10 Hz, so that they exhibit a high degree of reproducibility. The repetition rate of the entire device is very low: 1 or 2 shots per day. Whenever necessary, a pulse from the oscillator is picked up, undergoes some tailoring and is finally amplified up to a high energy. In the largest systems, several front ends are available. They deliver pulses with different shapes and contrast ratio. Pulse durations range from picoseconds (thousandth of a billionth of a second) to nanoseconds (billionth of a second). In all cases, these are very short times that one can hardly imagine. However, they can be illustrated by converting into distance travelled at the speed of light in vacuum (300000 km/s). Then 10 ps are 3 mm; 1 ns is 30 cm.

The output of the front end is a beam 1 mm in diameter. The diameter is to grow as the amplification proceeds. This can be done due to the natural divergence of the beam, a few milliradians. If it is not enough, telescopes are inserted in the chain at carefully chosen locations. They act

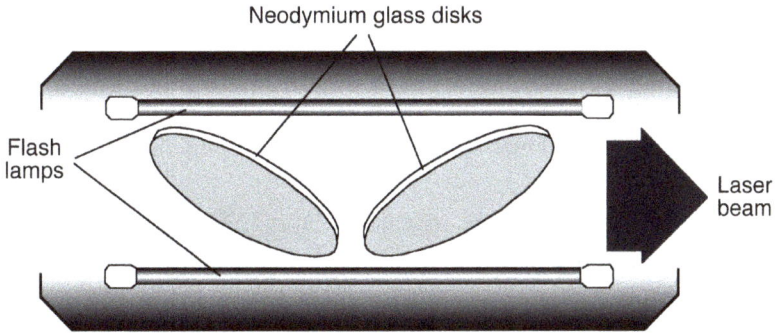

(a)

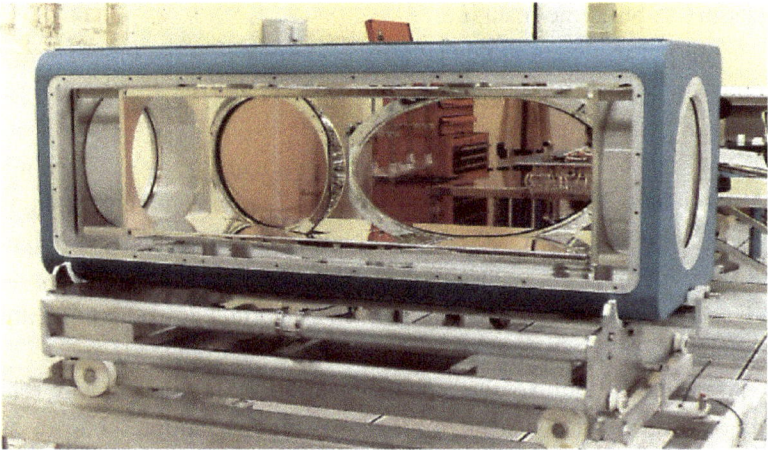

(b) (CEA document)

Figure 8.6. Disk amplifier in the 1980s: two antiparallel disks at Brewster angle. The pumping radiation and the laser beam enter the glass through the planar surfaces. Disk and flash lamps are mounted inside boxes fitted with reflective walls. (a) scheme of the amplifier; (b) an amplifying box of the Phébus laser (Limeil, France). The beam diameter was 75 cm.

as beam expanders. The final beam diameter is tens of centimetres, much larger than the beam duration converted into distance. The laser radiation is eventually packed inside a prolate cylinder propagating at the speed of light.

3. Megajoule lasers

After 40 years of technical advances, the construction of two megajoule lasers was planned. They will be operating during the 2010s, first in California, then in France.

In the US, the Livermore megajoule laser (NIF for National Ignition Facility) was completed in 2009 [1]. Table 8.2 displays a few data about this instrument.

The initial cost and the planned duration of the construction phase of NIF were underestimated by a large margin. Instead of 1.1 billion dollars and 6 years, the final price tag was 3.5 billion dollars, and 12 years were necessary to build the machine.

Light pulses were tailored (Figure 8.7) in view of the first experiments attempting at ignition that were performed in 2012 (National Ignition Campaign).

In France, at "Centre d'Etudes Scientifiques et Techniques d'Aquitaine" (CESTA), the megajoule laser (LMJ) is to be operational by 2014 [2]. However only 160 of the 240 beams will be actually installed in a first step. In this laser, high-power chains include innovative technologies such as four-pass amplifiers (Figure 8.8). The efficiency of amplification is thus increased. As a bonus, costs are consequently reduced.

As the construction of the Laser Mégajoule (LMJ) is under way, relevant technologies are tested in a single set of eight beamlines known as the "Ligne d'Intégration Laser" (Laser Integration Line or LIL). The basic amplifier for each beamline is made of nine neodymium glass plates. Eighteen laser plate cartridges are assembled to form an amplifying unit for the eight beamlines (Figure 8.9). Similar bundles are to be used in the Laser Mégajoule. After the final amplification, two subsets of four beams

Table 8.2. NIF megajoule laser (National Ignition Facility, LLNL).

Number of beams	192
Number of check points for beam quality	60 000
Distance travelled by the laser beams from oscillator to target	1 Km
Time interval during which all beams reach the target	30 ps
Energy on target (without frequency conversion)	1.8 MJ

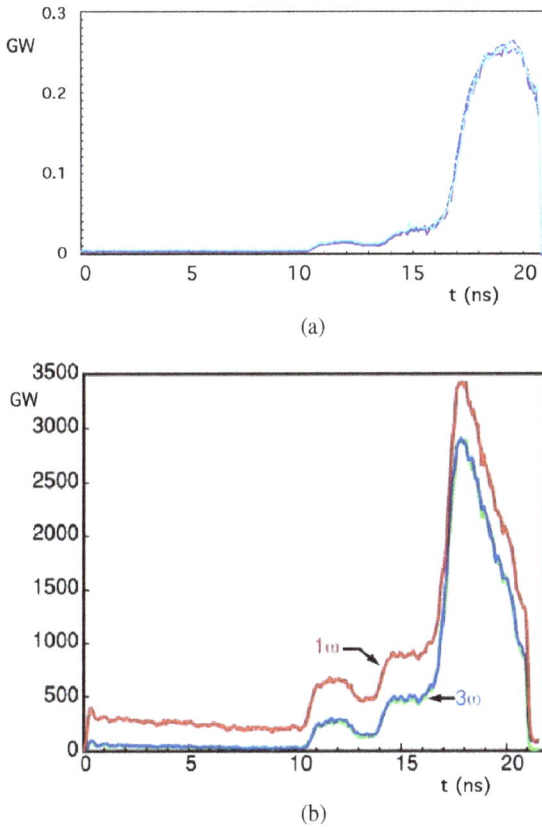

Figure 8.7. (after a LLNL document). Time histories of NIF laser pulses: (a) front-end output; (b) output of one among the 192 beams, without (1ω, red) or with (3ω, blue) frequency tripling; powers are given in GW.

dubbed "quadruplets" are transported via different optical paths (with the same length) to the interaction chamber.

The 240 beamlines (60 quadruplets) of the Laser Mégajoule are amplified within two large halls symmetrically located on either side of the interaction hall. The interaction hall is cylindrical with a diameter of 50 m. The overall length of the building is about 300 m.

As far as inertial fusion is concerned, state-of-the-art high-power neodymium glass lasers face two major setbacks. First, the efficiency is low, at most 1% in the devices required to deliver the necessary power for the thermonuclear ignition. Second, the technology precludes a convenient

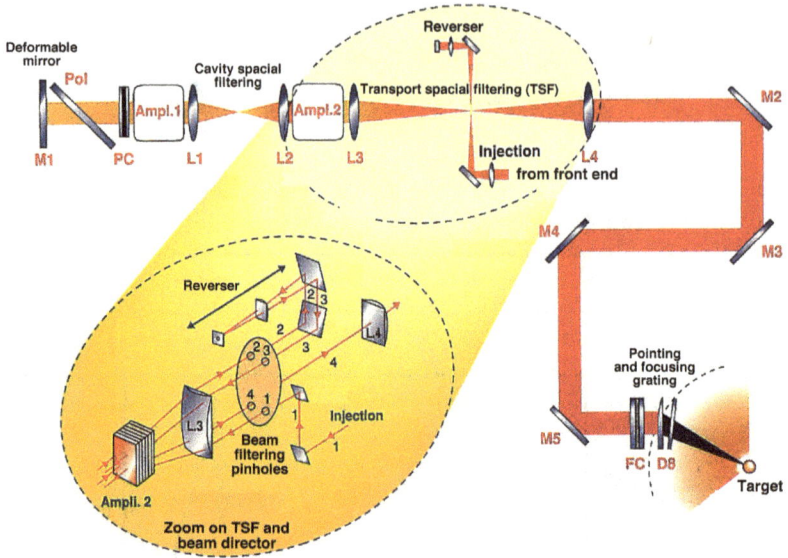

Figure 8.8. Layout of the four-pass high-power laser chain at Laser Mégajoule (LMJ). The main amplification stages are located inside an optical cavity fitted with a polarizer (Pol) and a Pockels cell (PC). When activated, the shutter rotates the polarisation plane of the light. The injected pulse, amplified after a first pass through ampl1, cavity spatial filtering and ampl2, is reflected back by the deformable mirror M1. The cell is activated. The pulse stays inside the cavity in which it makes a round trip (two more passes) before being reflected back again by M1. The cell is then disactivated. The pulse after a final pass through the amplifiers is transported towards frequency conversion (FC) and final focus. Note that the surface of M1 is adjustable in order to compensate defects in the light waveform, a technology borrowed from astronomical instrumentation (after a CEA document).

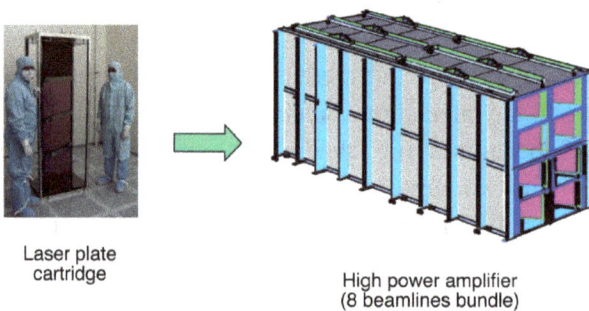

Laser plate
cartridge

High power amplifier
(8 beamlines bundle)

Figure 8.9. (after a CEA document). Amplifying unit in the LIL and the Laser Mégajoule. 4 neodymium glass plates are mounted in a single cartridge. An amplifier is made of 18 cartridges (72 plates) for an 8 beamlines bundle.

repetition rate. Consequently such lasers are not suited to a profitable energy source. New laser technologies are badly needed (Box 8.2).

Box 8.2. Towards high-efficiency solid state lasers

Flash lamp pumped neodymium glass lasers have a poor efficiency. An alternative method, optical pumping using diodes looks much more promising. Lots of semiconductor lasers emit radiation whose wavelength is tuned to a neodymium absorption line. The efficiency is then a few tens of percent. High repetition rates are achievable. Such devices proved effective at low energies. Reaching the megajoule size will require further developments that are to be conducted over a long period of time. At Livermore (LLNL), the Mercury program has been underway since 1996. In 2008, 50 J, 15 ns pulses were obtained with a 10 Hz repetition rate.

Laser radiation in the near ultraviolet can be obtained from eximers, i.e. transient molecules combining a noble gas and an alkaline. For instance, an electrical discharge through a high pressure mixture of fluorine and krypton produces excited KrF molecules. The efficiency of this electronic pumping is much higher than the efficiency of flash lamp pumping of neodymium glass. Although electronic pumping is used in many laser systems, no short pulse technology has been available so far. Research is still ongoing.

Other high-power lasers with a high efficiency such as carbon dioxide (CO_2) systems are not convenient for inertial fusion. For instance the wavelength of CO_2 lasers, $10\,\mu m$, is far too large for an efficient interaction with plasmas.

4. Beam to target

After amplification, the infrared laser light is to be converted to a higher frequency. Indeed, as it is well known after decades of research, obtaining a high pressure in the interaction plasma whilst staying rid of parametric instabilities, requires wavelengths in the near UV. In this spectral range, no laser source is available to deliver reliably high power pulses.

Now, high-efficiency frequency doubling occurs in KDP crystals (deuteriated acid potassium phosphate). The higher the incident power, the better is the efficiency up to 80%. Two-stage devices allow frequency tripling or quadrupling. Losses are comparatively low. Half the infrared power can be converted into near UV at the convenient wavelength $0.25\,\mu m$

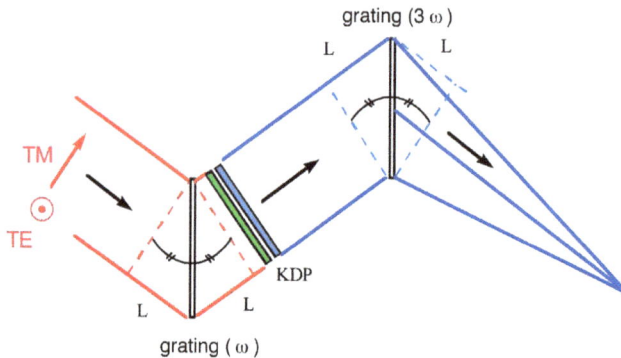

grating (3 ω)

L   L

TM

TE

KDP

L   L

grating (ω)

Figure 8.10. (CEA document). Combined frequency conversion and focusing devices as implemented at the Laser MégaJoule. The first grating (at ω) eliminates unwanted frequencies, whilst a double frequency converter delivers 3ω radiation to be filtered and focused by the second grating.

while keeping the pulse shape. Furthermore, since the frequency conversion processes are nonlinear, the peak to pedestal contrast is increased together with the coherency of the beam.

Now, growing big KDP crystals with good optical qualities is difficult and costly. The actual devices look like stained glasses. Several crystals are assembled within a metal frame. In the French Laser MégaJoule, the frequency conversion devices are associated with gratings that act as both frequency filters and focusing systems (Figure 8.10).

The light from a multi-beam, high-power laser is to be concentrated onto a single tiny target. Each beam passes through a high-precision vacuum-tight optical block located at a target chamber entrance port. In order to adapt to different target irradiation geometries, optical blocks are movable, but within narrow limits.

The spherical target chamber has the size of a hot air balloon (Figure 8.11). Figure 8.12 shows the interior of the target chamber at NIF with the target holder.

Ports fitted with optical blocks are located in such a way that homogeneous direct drive irradiation is made possible. The beams are to impinge onto the target simultaneously, i.e. within a few picoseconds time interval. The trajectories of the light follow long complicated paths (several hundred meters) that should be adjusted within a fraction of a millimetre. The many directional changes are obtained with mirrors mounted on rotating gimbals.

Figure 8.11. (CEA document). Laser MégaJoule: the spherical target chamber at its final location. Bundles of four laser beams are to enter the chamber through the square ports. Circular ports are for measurements and diagnostics.

Figure 8.12. Inside the target chamber at NIF (LLNL document). The target is at the centre of the white circle. Other converging holders are for diagnostics.

Table 8.3. Expected fusion performances with two different configurations of the Laser MégaJoule (L 1000 and L 1215) and a single configuration of the NIF laser.

	Laser MégaJoule (LMJ)		NIF
	L 1000	L 1215	NIF-PT
Laser energy (MJ)	1.45	1.6	1.3
Power (TW)	480	420	380
Temperature in the hohlraum (eV)	345	300	300
External radius (mm)	1000	1215	1110
Ablator thickness (mm)	180	175	160
DT thickness (mm)/DT mass (mg)	120/219	100/310	80/209
In-flight aspect ratio	25	32	37
Absorbed energy (kJ)	145	172	139
Convergence ratio	39	39	43
Fusion energy (MJ)	23.5	25.4	17.6
Gain (Fusion energy/laser energy)	16	15	13

Automated servo mechanisms are used for alignment and adjustment of optical paths. For instance at LIL, beam synchronisation is obtained within 0.03 ns.

Megajoule lasers are designed for irradiation of fusion capsules which contained DT filled microballoons. Sizes, aspect ratios, filling conditions and expected performances of microballoons to be actually implemented are given in Table 8.3. Note that stated laser energy and power are 25% larger than input values in the numerical simulations which produce the given fusion yield. To a certain extent uncertainties due to poorly mastered unwanted effects are thus accounted for. However, experiments conducted at NIF in 2012 within the National Ignition Campaign did not achieve ignition (see Box 8.3).

So far no fusion energy recovery has been planned in the present laser-driven inertial fusion programs. Experiments aim at the demonstration of small-size explosive thermonuclear burn. A small number of dedicated shots (typically 20) will be scheduled each year.

Box 8.3 Experiments towards ignition

At LLNL, the National Ignition Facility (NIF) was completed in 2009. In 2011–2012, a national ignition campaign was carried out. It attempted at central ignition in indirectly

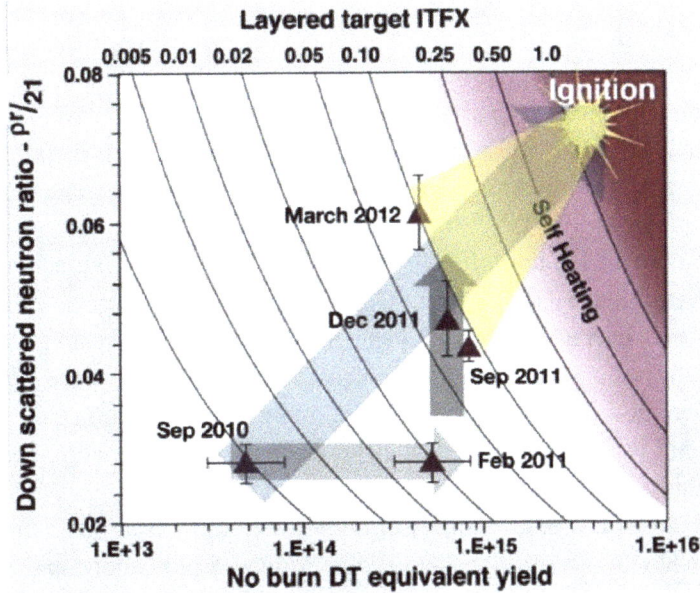

Figure 8.13. Parametric plot of the national ignition campaign.

driven cryogenic pellets [4]. After empirical adjustments of the laser pulse dynamics to the target design, hohlraum temperatures exceeding 300 eV were measured. However, the implosion velocity, slightly above 300 km/s, was lower than expected from numerical simulations. In the experiments performed with DT, the ion temperature ranged from 1.5 to less than 4.5 keV. Emitted neutrons were used as a diagnostic for both the fusion yield and since part of them undergo scattering before leaving the target, the areal density. The highest yield in terms of fusion energy was 2.5 kJ for a DT areal density of 1.3 g/cm^2. The internal energy of the central hot spot was thus found too low for an efficient ignition. Presumably, Rayleigh–Taylor hydrodynamic instabilities resulted in the mixing up of the D–T fusion fuel with heavier elements from structural materials. The main results were summarized using a parametric plot (Figure 8.13).

5. Spontaneous or triggered ignition?

At the megajoule level, the required compression for an efficient thermonuclear burn is likely to be obtained. However, at the end of the implosion process the temperature will be too low for a mass ignition. In Chapter 7, it was shown that central ignition due to shock wave convergence can be effective provided an almost perfect spherical symmetry prevails all along the compression stage. Then a central hot spot has a temperature above ignition conditions, a thermonuclear reaction driven heat wave propagates outwards and ends up in the sought for micro explosion. Such a violent burn occurring within a few picoseconds requires a very accurate tuning of the implosion.

This is indeed the difficult point. There is so far no evidence that the central ignition is to succeed. Many implosions of microballoons were performed during the last 25 years yielding abundant fusion reactions. Since the laser energy was far too small, the conditions were of no real significance for inertial fusion. Furthermore, the first attempts at ignition (National Ignition Campaign in 2012) once the NIF laser completed were unconclusive.

Meanwhile, the technology of high-power short laser pulses improved tremendously. During the late 1960s, trains of picosecond pulses from mode-locked oscillators were obtained, a first step towards ultrashort laser pulses. Afterwards, chirp (frequency modulation) was used for pulse compression down to femtosecond half widths. Amplification of picosecond and subpicosecond pulses was made possible after stretching by a pair of gratings that disperse the spectrum so that different frequencies follow optical paths of different lengths. Thus the flux density inside the active amplifying material remains below the damage threshold. After amplification, recompression needs a second pair of gratings that reverses the dispersion of the first one. This method proved effective to get output powers exceeding a terawatt (10^{12} W) [3]. The whole set up is compact enough and for this reason was dubbed T^3 (for Table-Top Terawatt). Recent advances pushed the power up to petawatts (10^{15} W), output of evidently less compact devices.

Whenever the light in high-power ultrashort laser pulses is concentrated by a lens, exceedingly high intensities up to 10^{20} Wcm^{-2} are obtained. At such intensities, interaction of radiation with matter opens a new field in fundamental research. A host of effects are observed either at the atomic or at the macroscopic level. Atoms and molecules are suddenly stripped of electrons. Free electrons oscillate and radiate at frequencies which are harmonics of the incident frequency. Furthermore, ponderomotive forces

Chapter 8. Big Drivers for Inertial Fusion

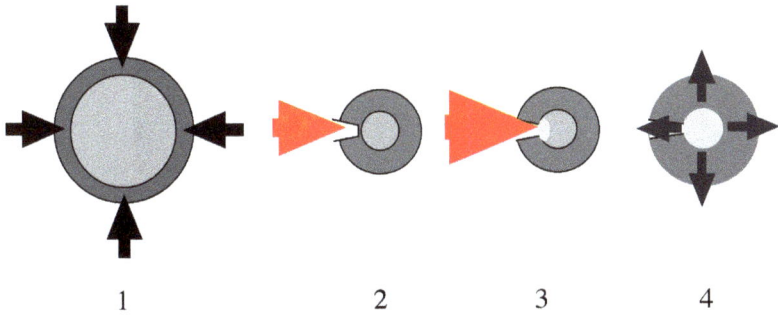

1   2   3   4

Figure 8.14. Laser-driven fast ignitor: 1) adiabatic compression of a microballoon; 2) self-focusing of an ultrashort laser pulse resulting in a hollow channel; 3) electrons are accelerated by another ultrashort pulse towards the DT in which their energy is absorbed thus creating a hot spot; 4) the thermonuclear burn invades the whole DT sphere.

expel electrons out of the focal region. Consequently, self focusing occurs. The laser beam stays concentrated far beyond the Rayleigh length at focus while drilling a low-density channel through the material. Electrons can be accelerated by another high-power laser pulse travelling along the channel. Hence the concept of a two steps fast ignitor [5] [6], according to the scheme presented on Figure 8.14.

A typical sequence for a fast ignitor is: first, a 100 ps high-power pulse is concentrated at the external surface of a compressed microballoon. At an intensity exceeding 10^{18} Wcm^{-2}, self focusing occurs, a channel is created through the external layers of the target and extends close enough to the thermonuclear fuel. Then a 5 ps laser pulse is focused at the channel entrance with an intensity 10^{20} Wcm^{-2}. Electrons are accelerated and penetrate the DT mixture in which they are absorbed. A hot spot is thus induced in order to start the thermonuclear burn.

In an alternative set up, the microballoon is fitted with a metal cone that serves as a guide for the laser pulses and enhances beaming of the accelerated electrons (Figure 8.15).

Initially, in fast ignitor studies, it was assumed that a comparatively modest laser energy would trigger the ignition of compressed DT. However, detailed investigations ended in a different conclusion. Fast ignition is efficient only if the laser energy is in the rage of tens of kilojoules, an energy to be accounted for in the overall balance of the laser-driven inertial fusion process. Consequently, in the European project HYPER [7], the energy for

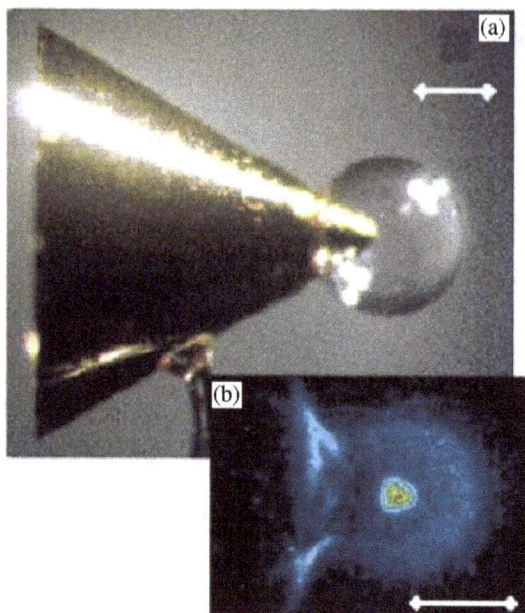

Figure 8.15. Fast ignitor with guiding cone (ILE Osaka documents). (a) Guiding cone fitted on a glass microballoon; (b) autoradiography during laser interaction with such a target.

the laser-driven compression is 200 kJ delivered in 5 ns whilst fast ignition requires a 70 kJ, 10 ps laser. The final design is due by 2013. The facility could be operational by 2020.

6. An alternative solution: particle beams

High-power lasers are energy sources whose power and pulse duration are well suited to inertial fusion. However, a major setback is the poor efficiency of flashlamp-pumped neodymium-glass lasers. On the contrary, particle beams can be efficiently produced: more than 50% of the input energy goes into the energy of the accelerated particles. High-energy particle beams are an alternative driver for inertial fusion, direct or indirect.

The first attempts at particle-driven fusion used electrons. In the 1970s at Sandia Laboratories (Albuquerque, USA) and at the Kurchatov institute (Moscow, then Soviet Union), high intensity electron beam generators aimed at fusion were built. However, it was soon realized that electrons are not the best possible particles. First, even in relativistic conditions the particle density inside the beam is space charge limited. Even with strong focusing, high current densities cannot be achievable unless the energy of individual electrons is considerably increased. Then the range of electrons in solids is far too long and the required energy densities (pressures) in matter for inertial fusion are not obtained.

Due to mass and charge effects, ion beam are free of such drawbacks. The important point is then the stopping range in matter, shorter in the case of a higher mass particle. Now consider ions from heavy elements: high charges also favour the slowing down of ions inside a material. However beams of highly charged ions are not needed since whatever the charge of particles impinging onto a target, they are immediately stripped of the remaining electrons. Transportation of singly charged ion beams that minimize space charge effects is then a definite advantage. Consequently, heavy ion beams look most interesting for inertial fusion [8]. In order to illustrate the point, Figure 8.16 shows for different species, the calculated stopping ranges of aluminium as functions of ion energy.

Since the hot dense matter created by ionic impact is a strong radiator, heavy ions are also convenient for indirect drive inertial fusion. On Figure 8.17, two examples of capsules are schematically represented. In the basic two-beam design, converters are located at both ends of a cylindrical

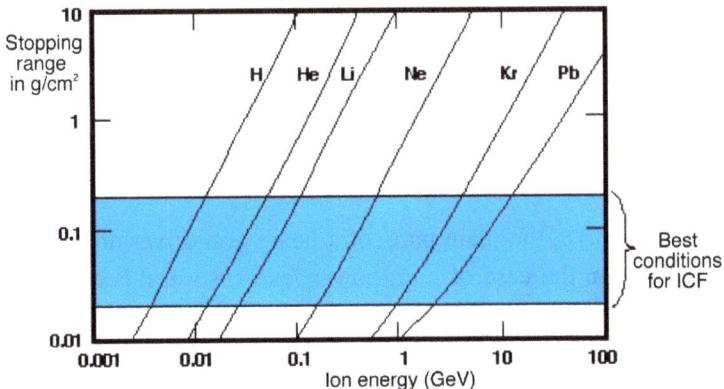

Figure 8.16. Stopping ranges in a dense aluminium plasma (0.2 g/cm^3 at temperature 200 eV) calculated in g/cm^2 for different ion species. Energy deposition in a areal density about 0.1 g/cm^2 is convenient for starting an implosion process.

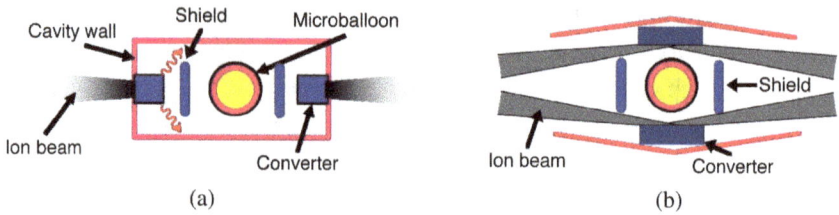

Figure 8.17. Indirect drive with ion beams: (a) basic scheme of a hohlraum with end plug converters; (b) "close coupled" target under study at Livermore.

cavity (hohlraum). The ion beam energy transforms matter into a radiating dense plasma with a temperature of a few hundreds eV. Recently, another design named "close coupled" was proposed. Surrounding a microballoon, the converter is cylindrical and many ion beams impinge onto it with oblique incidence. A better efficiency is thus expected. In both designs, shields contribute to the homogeneity and isotropy of the thermal radiation.

Projects dealing with ion-beam-driven inertial confinement fusion had an impact on accelerator physics and technology. Contrary to up to date colliders aimed at particle physics (LHC at CERN) that use nanocoulomb proton bunches in the TeV range, inertial fusion is best induced with 10^{-3} coulomb ion bunches accelerated up to GeV. For instance, 3 MJ represent 10^{15} 10 GeV Bismuth ions. Specific methods and devices are thus required [8]. Some are already being tested in the laboratory [9] whilst small scale heavy ion interaction with matter are under way.

Heavy ion acceleration is routinely performed at GANIL (National Large Heavy Ions Accelerator) in France and at GSI (Society for Heavy Ion Research) in Germany. However, these facilities are not suited to heavy ion fusion.

In Figure 8.18, the main parts of a heavy ion driver are symbolically presented. As in the case of high-power lasers, several beams in parallel are needed. Each one starts with a laser-triggered pulsed single-charge ion source. Ion packets are accelerated first along a sequence of radio frequency quadrupoles and drift tubes linear accelerators (DT Linacs). Then they are injected into the main accelerator, another Linac, in which they acquire a 10 GeV energy. Accumulation and phasing of ion packets take place in a system of storage rings. Pulse compression is obtained in a radio

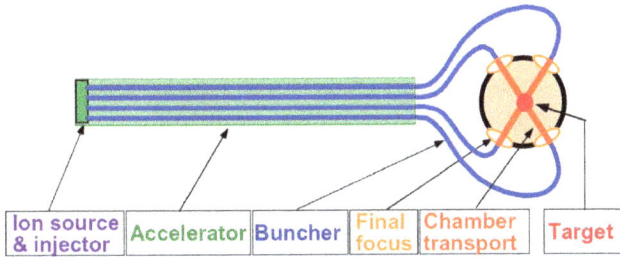

Figure 8.18. Scheme of an accelerator system for heavy-ions driven inertial fusion (after J.L. Vay, LBNL).

Table 8.4. Expected performances of a heavy ion driver for inertial fusion.

Ion energy	3 to 10 GeV
Total energy from the driver	3 to 5 MJ
Power	400 TW
Total current	40 kA
Width of the final pulse	6 to 10 ns
Focal spot diameter	3 mm

frequency "buncher" before final transport and focusing. Pulse compression and focusing are the most challenging points.

Detailed studies about each one of these stages show that a facility fulfilling the requirements stated in Table 8.4 is indeed feasible.

From source to target, heavy ions travel along very long paths for kilometres. The facilities to be built are as gigantic and costly as high-power lasers.

7. The "Z-pinch" is back

In an indirect-drive scheme the actual implosion driver is the thermal radiation source, whatever is the origin of it. As it was shown in the previous sections, laser light or particle beam interaction with solids create convenient radiation sources. How about electrical discharges?

High-power current pulses are a commonplace technology. Methods for concentrating such pulses and transferring their energy into a small volume of matter are well known. For instance a collapsing high-intensity plasma column is a strong radiation emitter. The heavier the elements in the plasma, the more intense is the emitted radiation.

Although it is not suited for directly creating fusion condition, the Z-pinch has been under study for decades as a simple and comparatively cheap plasma source. Now, it can be designed in such a way as to deliver short intense X-ray pulses. To this end, a gas is not the best initial state. It is far better to start from a metallic hollow cylinder made of either a liner or a wire array. The leading edge of the current pulse explodes the metal transforming it into a plasma sheet to be projected towards the axis by electromagnetic forces. Heating occurs during the cylindrical implosion. The final temperature can be made large enough for an efficient radiation source.

The efficiency of the process depends upon the choice of materials to be transformed into plasma, the geometry of the device, and the coupling with the pulse generator. Combining a radiation emitting Z-pinch with a hohlraum aimed at imploding microballoons looks quite sensible. Specific designs have been investigated. Figure 8.19 is an example from Sandia laboratories [10]. In this set up, two radiation sources are located symmetrically on both sides of a microballoon. In the primary holhraums, wire arrays were chosen. Indeed liners are more sensitive to hydrodynamic instabilities during the implosion. Wires are made of heavy metal: tungsten is an appropriate choice. The current pulse blows off the wires to form a plasma sheet set into centripetal motion thanks to the electromagnetic force. The imploding plasma sheet impacts an axial target made of low density foam in order to stabilize a hot dense radiating medium.

Very high pulsed power is involved in Z-pinch devices whose radiative performances are reported on Figure 8.20. The dramatic progress visible between 1995 and 2000 is due to the implementation of the Z generator (data in Table 8.5) the latest facility implemented at Sandia. An artist's view of the machine is presented in Figure 8.21.

In the 2000s, the Z machine received many improvements dealing with the reproducibility and the tailoring of the pulses. State-of-the-art performances are a 20 MA peak current reproducible within a 1% margin and a 100 ns FWHM[1] pulse. 50 GW of electric power are fed into the wire

---

[1]Full Width Half Maximum.

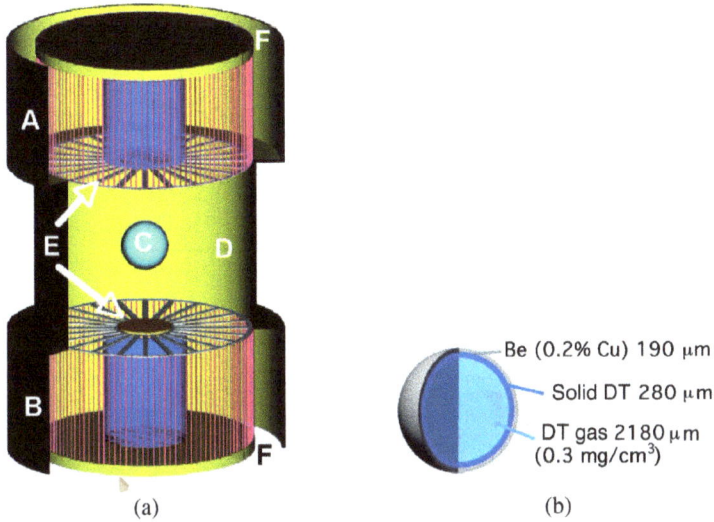

(a)   (b)

Figure 8.19. Conceptual design of a hohlraum irradiated by a double Z-pinch (Sandia document). (a) Setup: A and B are primary hohlraums with Z-pinch wire arrays and axial foam targets: blue cylinders; D, secondary hohlraum around the microballoon C; E are axial shields opaque to the direct radiation of imploded wire arrays; F are conducting plates. (b) Cutaway of a high gain microballoon.

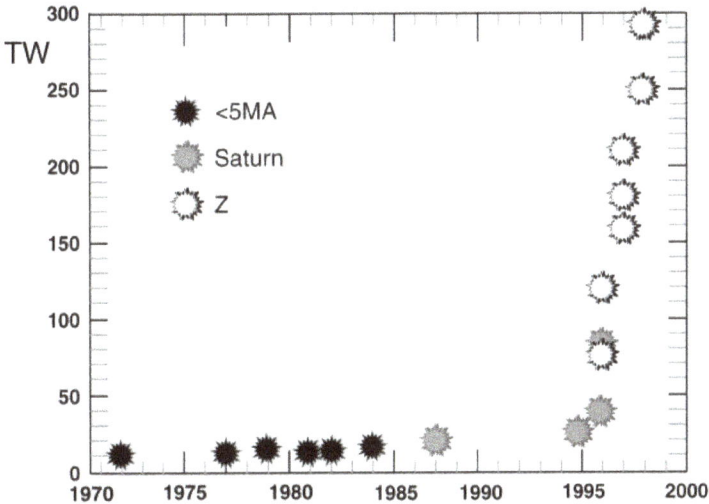

Figure 8.20. Time history of X-ray power (in terawatts) emitted per shot in Z-pinch devices. Symbols refer to successive generation of pulsed power generators built at Sandia laboratories (New Mexico, U.S.A.).

Table 8.5.	The Z machine.
Overall diameter	33 m
Energy stored in capacitor banks	20 MJ
Energy received by the target	5 MJ
Peak current	>20 MA
X-ray energy	2.7 MJ
X-ray power	350 TW

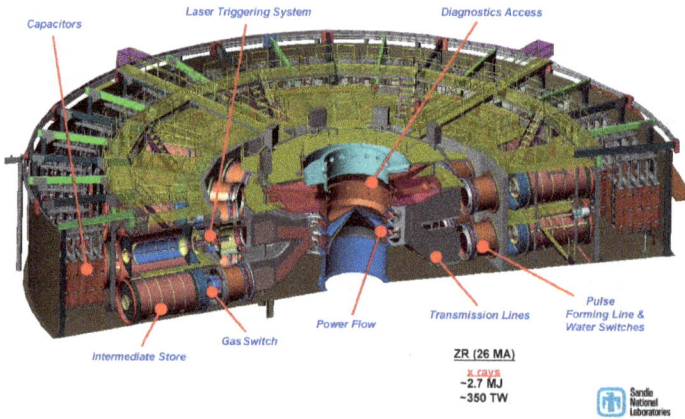

Figure 8.21. An artist's view of the Z machine operational in 2007 (Sandia document). The outer circle houses capacitor banks, the primary storage for the energy to be delivered in the high-power pulse. Capacitors are charged in parallel and discharged in series (Marx generator). Energy is transferred first to an intermediate storage. Then through a gas switch, it is injected into a pulse forming line. Finally, the current pulse is transported to the target.

array. The imploded axial plasma emits 300 TW of X-ray radiation in a 5–10 ns FWHM pulse (Figure 8.22).

Hydrodynamic instabilities are still plaguing the radiative emission of imploded wire arrays. However an increase of 50% was obtained using a double coaxial layer of metallic wires, blown off by parallel currents. The corresponding radiative energy exceeds 2 MJ.

Usually radiative temperatures inside the hohlraum are of order 200 eV ($\approx$ 2 million degrees Celsius). In some experiments, temperatures around 2 billion degrees (200 keV) were observed [11]. This remarkable result could be of interest for the ignition of thermonuclear fuels other than DT.

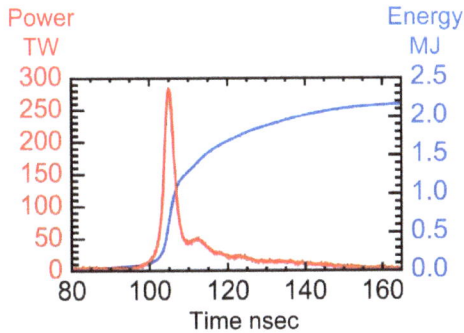

Figure 8.22. Radiative power (in red) and energy (MJ, in blue) emitted in the X-ray spectral range by the Z machine. Results obtained in 2008 (Sandia document).

While the Z machine is running, research is still being carried out on pulsed power: improved efficiency of current generators, optimized tailored pulses, and repetition rates. Within a joint program of Sandia laboratories and Russian laboratories including the Kurchatov institute in Moscow, new pulse shaping system dubbed LTD (Linear Transformer Driver) and a new final beam transport line RTL (Recyclable Transmission Line) are being tested.

Z-pinch machines aimed at inertial fusion are big facilities. Furthermore numerical simulations of the dynamical behaviour of the imploding wire arrays require the most powerful computers available to date. As in the case of high-power lasers and ion beams, the Z-pinch approach to ICF belongs to the realm of big science.

References

1. https://lasers.llnl.gov/.
2. http://www-lmj.cea.fr/.
3. D. Strickland and G. Mourou, *Optics Comm.* **56**, (1985) 219.
4. O. L. Landen *et al.*, *Plasma Phys. Control. Fusion* **54** (2012) 124026; see also papers appearing in *Physics of Plasmas*, Vol. 20, special issue 54th Annual Meeting of the APS Division of Plasma Physics, May 2013.
5. M. Tabak, J. Hammer, M. E. Glinsky, W. L. Kruer, S. C. Wilks, J. Woodworth, E. M. Campbell, M. D. Perry and R. J. Mason, *Phys. Plasmas* **1** (1994) 1626.
6. Special issue on fast ignition, *Fusion Science and Technology* **49** (2006).
7. http://www.hiper-laser.org/.

8. B. G. Logan *et al.*, *Nuclear Instruments and Methods* **A577** (2007) 1.
9. http://hif.lbl.gov/tutorial/tutorial.html.
10. http://www.sandia.gov/z-machine/.
11. M. G. Haines, P. D. LePell, C. A. Coverdale, B. Jones, C. Dewney and J. P. Apruzese, *Phys. Rev. Lett.* **95** (2006) 075003.

Off the Main Trails

In Latin ITER means the way. Which one? Fast track or dead end? Actually, no one knows whether or not ITER will lead to a profitable electricity production in a further generation of up-scaled tokamaks. Concentrating most of the available subsidies and manpower on a single project for magnetic fusion is a questionable policy. A similar doubt applies to the emphasis put on high-power lasers in the realm of inertial fusion.

Alternative proposals such as fusion–fission hybrids, other confinement schemes, muon catalysis, fusion reactions other than DT, etc., are all worth serious investigations. The corresponding research programs are carried out on a modest pace compared to ITER and megajoule lasers.

1. Fissile blankets

In a position paper issued in 2007 [1], the European Physical Society stated: "*the nuclear option should mean consideration of energy production by both fission and fusion processes*". Since DT fusion produces a lot of 14 MeV neutrons, why not combine the two? Fusion reactions could be used as intense neutron sources and conceptually replace the spallation source irradiating a sub-critical fissile assembly in Accelerator Driven Systems (ADS) [2].

The oldest proposal (1951) for a fusion–fission hybrid is referred to in Andrei D. Sakharov's *Memoirs* [3]. It deals with the use DT neutrons for fissile elements production. 20 years later, hybrids were under study [4]. By the end of the 20th century, it was not a very popular idea. Indeed with hybrids, fusion can no longer be advertised as "clean" nuclear energy. Highly radioactive waste (fission products and actinides) would be released, and proliferation threats would still be a concern.

In the blanket of a pure fusion reactor, 14 MeV neutrons slow down and their energy (with possibly a slight gain g) is converted into heat to be transferred to a circulating fluid. Interacting with lithium, neutrons will also be used for tritium breeding. Of course, neutrons could do more should the blanket incorporate a sub-critical array of fissile materials.

Entirely controlled by the fusion reaction, neutron multiplication would result in a high gain, the main advantage of a fusion–fission hybrid. Indeed a DT fusion reaction releases about 18 MeV to be compared to 200 MeV from a fission induced by a neutron. Moreover, fission cross sections with 14 MeV neutrons are large for any Thorium isotope (including 232) and any Uranium isotope (including 238) as well. 4 or 5 secondary neutrons are emitted in the process. Within a fissile blanket, a gain g with respect to the fusion energy can be rather easily made larger than 10. Adjusting the effective multiplication coefficient, a high g of 150 is foreseeable while tritium regeneration (one T nucleus per incident neutron) is efficiently secured. Hybrids could thus take part in making the most of fissile resources, as it is also expected from fourth generation breeder nuclear reactors.

Both functions of the blanket, a high gain in energy and an efficient tritium breeding, are to be optimized. Figure 9.1 shows a cutaway of a possible design [5].

As far as safety is concerned, since the blanket is sub-critical, no runaway of the fission reaction is to occur. However, some problems of fission reactors still remain. In case of sudden breakdown, the residual heat power due to radioactive products should be evacuated. Furthermore, the question of radioactive wastes is open with a slight difference: as it will be seen later fusion–fission hybrids could contribute to waste elimination.

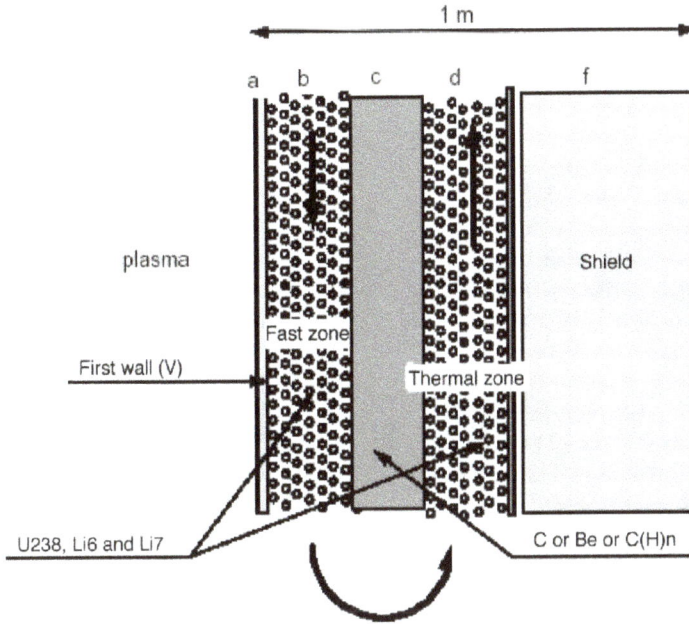

Figure 9.1. Example of an optimized fissile blanket: a) first wall in vanadium; b) first active zone incorporating uranium and lithium (14 MeV neutrons); c) neutron moderator layer; d) second active zone (slow neutrons); f) shield (40% of the total thickness) to absorb the residual particle flux. Both active zones are cooled by circulating water (arrows).

2. Hybrids with magnetic or inertial confinement

In a tokamak surrounded by a fissile blanket, there is no reason for a very high fusion gain Q, i.e. the ratio between fusion power and auxiliary heating power. Q < 10 could be enough for an economically profitable electricity generation. However the concept is still to be successfully tested.

For instance the modular blanket of ITER could be used as a preliminary test bench. Although the project does not include so far the possibility, a module could incorporate depleted uranium with a design allowing an increased heat extraction. At ITER size, a fissile blanket could produce 4 GW of thermal energy, the same amount as in the case of third-generation nuclear fission reactors.

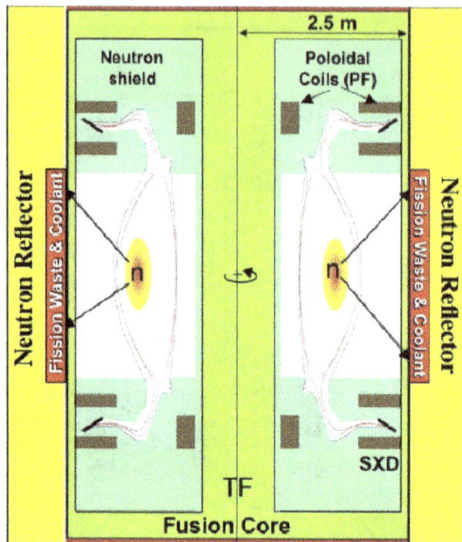

Figure 9.2. Hybrid fusion fission project with a compact torus. The toroidal field is generated by the axial conducting rod. The blanket is cylindrical. Fission waste stands for used fission fuel: depleted uranium and actinides.

The performances of the fusion device in a hybrid reactor are well below the requirements of a pure fusion machine. Magnetic configurations other than tokamaks presently discarded or surviving through marginal programs, could accordingly arouse renewed interest. Mirror machines or compact tori (see Chapter 5) are reconsidered in this context. This is the case in America for simple compact tori yielding moderate fusion power. Figure 9.2 shows a typical example of such a reactor [6].

In this design, the major radius is 1.35 m and the aspect ration is 1.8 with a high elongation. The ratio β is non standard: 15 to 18%. The 14 MA current is very large for the size of the plasma ring. The fusion power in H mode would be 100 MW with a fusion gain Q around 2. All these value could be obtained with 2010 technologies. Among the unknowns, the operation is expected to be in a quasi-continuous mode implying 50 MW of current drive generation.

State-of-the-art technologies would be used for the blanket: liquid sodium as a coolant, fuels elements similar to fast neutron reactors. Fissile materials would be spent fuels from pressurized water reactors (PWR).

Table 9.1. Comparison of a pure inertial fusion reactor with a laser-driven hybrid.

	Pure fusion	Hybrid
Laser energy (MJ)	3 to 5	1.5
Repetition rate (Hz)	15	15
Fusion energy (MJ)	300 to 500	20 to 30
Multiplication factor of the blanket	1	10
Overall gain	100	100 to 300
Total thermal power (GW)	3	2 to 3

The neutron flux at the first wall would be at most $4 \times 10^{13}\,\mathrm{cm}^{-2}\mathrm{s}^{-1}$. Tritium regeneration would be achieved with lithium titanate embedded in the blanket. The total thermal power, 2 to 4 GW, could be lowered in order to ensure a better elimination of transuranides.

Projects of fusion–fission hybrids are also elaborated starting from inertial confinement devices [7]. Recent studies combine energy production and nuclear waste incineration. At Livermore, the Laser Inertial Fusion–Fission Energy (LIFE) program is an offspring of NIF [8]. A favourable energy balance is obtained with 20 to 25 MJ only of fusion energy per micro-explosion (Table 9.1).

20 MJ of fusion are expected with upgraded megajoule lasers. Unfortunately the efficiency and the repetition rate of state-of-the-art high-power neodymium glass lasers are too low in any case. The emphasis of future development is to be put on increasing the efficiency up to tens of percents and achieving repetition rates of a few hertz. These are precicely the objectives of the Mercury project (see Chapter 8).

In LIFE (Figure 9.3), the mass of the blanket is 40 tonnes. The first wall is made of steel that needs cooling. The next layer includes beryllium spheres for neutron multiplication with n–2n reactions. Neutron energy is thus lowered for a better fission rate in the following sub-critical assembly. Fissile material are immersed in a molten salt called FLIBE ($2LiF + BeF_2$). This 640°C liquid is both a coolant and a tritium regenerator. The blanket could accommodate several kinds of fissile materials: depleted uranium, spent nuclear fuels, plutonium from weapon decommissioning and radioactive waste. The fission energy would be comparable to nuclear reactors in power plants operating today.

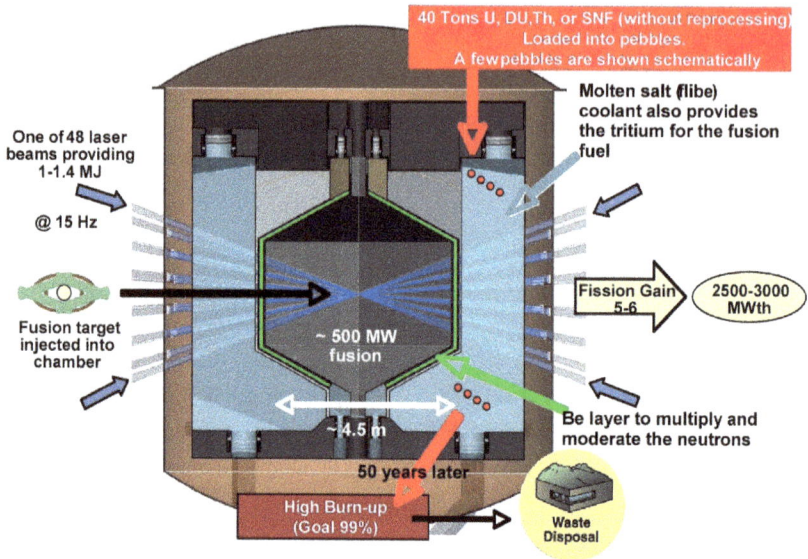

Figure 9.3. General scheme of a LIFE hybrid reactor. The fusion driver is a megajoule laser. Within the blanket, fissile elements could be made of natural thorium, of natural or depleted uranium, of spent nuclear fuels reprocessed or not (LLNL document).

The LIFE-type hybrid is a once through system. Fissile materials stay for the whole life of the system: some 50 years including 35 years at full power (Figure 9.4).

Drivers other than lasers are also considered for inertial fusion hybrids. Under the explicit name InZinerator, Sandia laboratories are developing a project aimed primarily at eliminating nuclear waste (Figure 9.5). An upgraded Z machine with a 0.1 Hz repetition rate is used to power the fusion part.

The fission part has many similarities with the core of an ADS. Within the blanket, a neutron absorbing liquid, a mixture of lithium fluoride and actinide fluorides, flows through tubes immersed in 650°C molten lead. Differences with the LIFE project stem from the lower repetition rate. Indeed a 0.1 Hz repetition rate is a maximum for state-of-the-art Z-pinch technologies. The fissile assembly should be close to critical with an effective neutron multiplication factor of 0.97. Then the energy multiplication factor is 150 for a 3 GW thermal production amplifying a

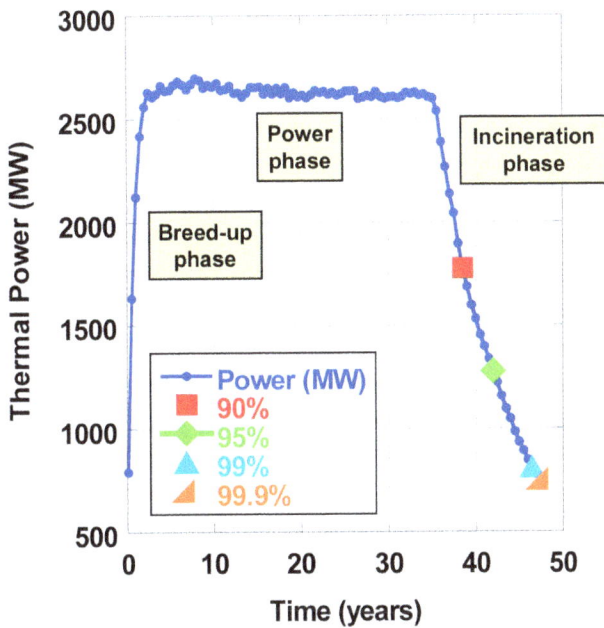

Figure 9.4. Expected life cycle of a LIFE-type fusion–fission hybrid reactor. Percentages refer to the fraction of initial fuel burned up.

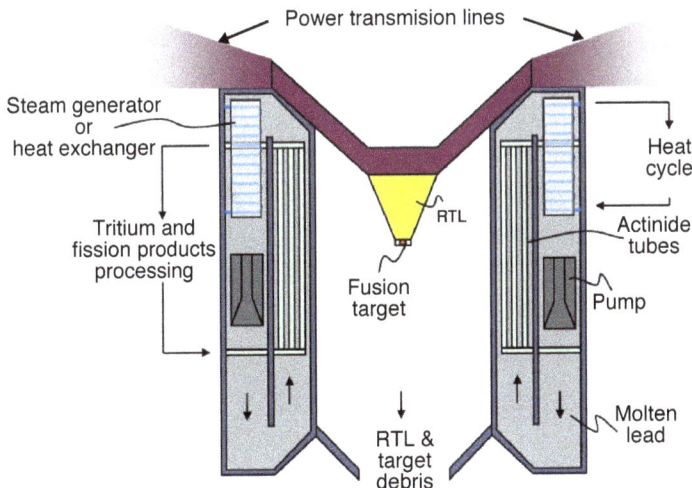

Figure 9.5. Scheme of an InZinerator. Pulsed power goes to the fusion capsule thanks to a recyclable transmission line (RTL). The micro-explosion occurs in an argon atmosphere seeded with aerosols for shock wave damping.

Table 9.2. Main parameters of an InZinerator.

	InZinerator
Energy released by the fusion capsule	200 MJ
Repetition rate	0.1 Hz
Transmutation rate	1 200 kg/year
Reactive mixture	$(LiF)_2$-$ActF_3$
Effective neutron multiplication coefficient k_{eff}	0.97
Energy multiplication factor	150
Thermal power of the reactor	3 GW (1 GW electric)
Coolant	Liquid lead
Working temperature	650°C
Overall diameter	6.4 m
First wall thickness	5 cm

200 MJ only fusion energy per shot. The main parameters of the device are given in Table 9.2.

Existing hybrid projects rely on the success of magnetic or inertial confinement fusion experiments to be performed during the coming years. Besides scientific and technological hurdles, a sociological obstacle is to be overcome. Indeed, communities dealing with nuclear engineering on the one hand and with fusion on the other hand are loosely connected. Researches are independently carried out on fourth-generation fission reactor, magnetic fusion, and inertial fusion. Weaving closer links would be desirable in order to complete hybrid projects that could be integrated in the development of nuclear technologies during the 21st century.

In a recent review, [9] a progressive implementation of hybrids is considered: first as incinerators, then as fissile elements makers, eventually for power production. To this end, more R&D efforts are to be devoted to nuclear technologies and fusion engineering.

3. Cold fusion catalyzed by muons

In the spring of 1989, cold fusion by heavy water electrolysis made the headlines. It turned out that the overadvertized results were definitely irreproducible [10]. Such processes are not to be dealt with here. However, getting rid of the thermonuclear regime which implies a 10^8 K plasma is still worth considering. Catalysis using muons is an alternative and more promising way of avoiding high temperatures [11].

Muons are heavy leptons. A muon is unstable and, with a 2.2×10^{-6} s half life, decays by weak interaction into an electron and two neutrinos. Muons are commonplace in high energy physics. They are encountered in cosmic rays and in laboratory experiments performed with particle accelerators. The so called "muon factories" imply dedicated accelerators.

Box 9.1 Muon production

In a muon factory, protons are accelerated up to 1 GeV. Impinging onto a heavy target, the protons create charged pions (symbol $\pi^{\pm}$) via either one of the reaction with nucleons: $p + p \to p + p + \pi^{+} + \pi^{-}$ or $p + n \to p + p + \pi^{-}$. Charged pions are unstable with respect to the weak interaction and decay with half-life 2.6×10^{-8} s into a muon with the same sign and a mesic neutrino. For instance: $\pi^{-} \Rightarrow \mu^{-} + \nu_{\mu}$.

Muons and electrons have the same electric charge. The simplest molecular ion is H_2^{+} with a single orbiting electron. Now the electron can be replaced by a muon. Since the mass of the latter is 200 times the mass of the former, the distance of the two nuclei is reduced by a factor of a thousand to 5×10^{-13} m. At such a small distance, the nuclear wave functions overlap. If instead of two protons the nuclei are a deuteron and a triton, D–T fusion occurs immediately. Now, a way does exist to create a D–Tμ mesic molecular ion.

When a muon collides with the molecules of a gaseous DT target, it extracts either a deuteron or a triton forming a mesic atom. A further charge exchange collision creates a mesic molecular ion. D–Tμ formation is enhanced by a resonant process:

$$[T\mu] + D_2 \to [[D - T\mu]^{*}D2e]^{*}$$

where the star * denotes an excited state. Actually in the hydrogen like D–(D–Tμ) molecule, the mesic ion replaces one of the D according to the scheme presented in Figure 9.6.

Once the fusion reaction is completed, the high-energy helium nucleus and neutron are ejected from the molecular assembly. The muon escapes and possibly enters a new cycle of D–Tμ formation and the process can be repeated. Accounting for the energy spent in muon production, the energy balance is positive whenever a single muon participates in more

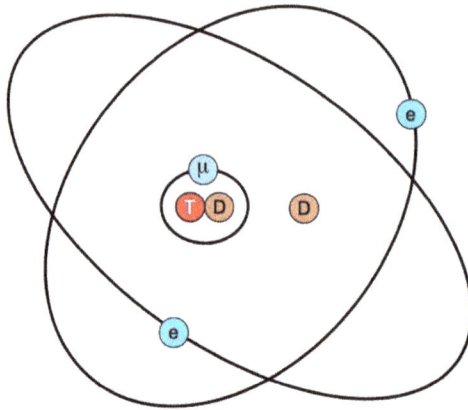

Figure 9.6. In this hydrogen like molecule, one of the deuterons is replaced by a mesic ion. Distances between D and T nuclei in the mesic ion and between this ion and the other deuteron are not drawn to scale.

than 250 catalyzed fusion cycles. Sticking, i.e. the muons stays bound to the helium nucleus, is a detrimental mechanism with a low but poorly known probability.

It was demonstrated experimentally that the number of fusion per muon increases with the target density. Up to 150 fusions per muon were observed contrary to early theories which predicted in the 1980s a maximum of only 80. 350 fusions per muon are expected in future experiments not planned yet. Muon catalysed fusion is still attractive. However, since it needs a muon factory, it remains in the field of costly big science.

4. Fuels beyond deuterium–tritium

In most of fusion research, only the D–T reaction is considered. This is indeed the easiest reaction to be implemented in the experiments of the early 21st century. Tritium production is then a mandatory part of the projects. The complexities of tritium handling were tested at JET and ITER will benefit from it.

Other fusion reactions that do not involve tritium are presented in Table 9.3.

Now, as shown in Figure 9.7, at a given temperature, thermonuclear reaction rates of other reactions are smaller than that of D–T. Implementing

Table 9.3. Tritium-free fusion reactions.

$$D + {}^3He \rightarrow {}^4He(3.7\,MeV) + H(14.7\,MeV)$$
$$P + {}^{11}B \rightarrow 3\,{}^4He(8.68\,MeV)$$
$${}^3He + {}^3He \rightarrow 2p + {}^4He(12.86\,MeV)$$

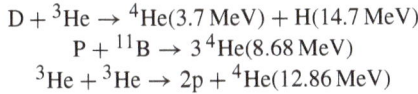

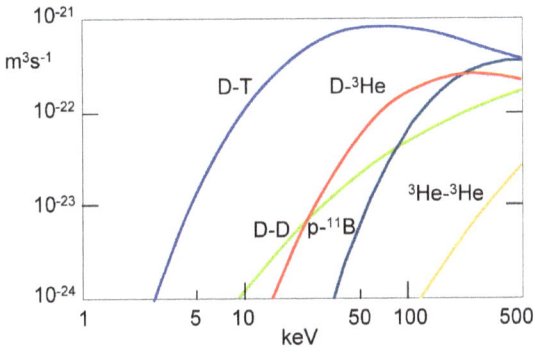

Figure 9.7. Compared reaction rates in the thermonuclear regime. The D–^{3}He reaction rate at 70 keV is the same as the D–T rate at 10 keV. p–^{11}B reaction rates include enhancement due to resonances in the sequential decay of the compound nucleus via transient ^{8}Be [12].

such reactions requires higher temperatures, thus imposing a further challenge.

Putting aside D–D reactions whose energy yield is too low, the first tritium-free reaction to be considered is D–^{3}He. Its reaction rate at 70 keV is the same as the D–T rate at 10 keV. At the microscopic level, the energy released in a single event is the same. No neutron is emitted in the case of D–^{3}He, all reaction products are electrically charged. Helium nuclei will stay in the plasma whilst high-energy (almost 15 MeV) protons escape. Appropriate electrodes and magnetic field could favour proton beaming into high-intensity electric currents, a convenient starting point for direct energy conversion.

In most fusion projects, the neutron energy is converted into heat. Using sophisticated technologies for eventually heating water looks unsatisfactory. On the contrary, direct energy conversion is an attractive concept that Inertial Electrostatic Confinement (IEC) would render effective. In this scheme, accelerated ions at the bottom of a potential well are trapped.

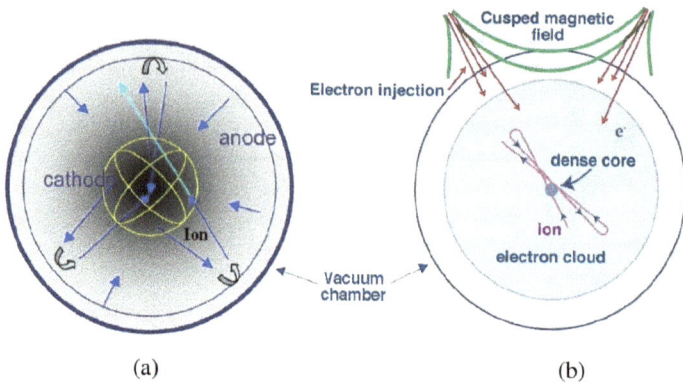

Two arrangements for an electrostatic confinement: a) a negatively polarized spherical grid (yellow) serves as a cathode inside a concentric spherical grounded anode; b) electron cloud magnetically confined inside a cusped magnetic field. In both cases, ions are injected from external sources.

The creation of a potential well can be obtained with electrodes: grids. An alternative solution uses an electron cloud confined inside a multipolar magnetic field. Both options are represented in Figure 9.8. In electrostatic confinement, since the ions are accelerated, their energy might correspond to the maximum cross section. Reaction rates are therefore higher than in the thermonuclear regime. Furthermore, the spherical geometry helps ion concentration near the centre.

When fusion reactions occur in a D–^{3}He mixture, 14.7 MeV protons from the D–^{3}He reaction and 3 MeV protons from the D–D reactions are detected. 2.45 MeV neutrons from the D–D reaction are used as a complementary diagnostic. Experiments were performed in Japan and USA (University of Wisconsin) with varying voltage differences between grids [13]. Applying a 50 then a 160 kV difference, the number of D–D reactions was multiplied by 30 whilst the number of D–^{3}He reactions was multiplied by 10^5. The source of reaction products could be localized. D–D reactions are produced preferentially in the central plasma. On the contrary, most D–^{3}He reactions occur on the grids. Indeed grids absorb ions from the plasma and are submitted to high-energy ion bombardment. Although interesting results were obtained on the properties of electrostatic confinement and on the subsequent fusion

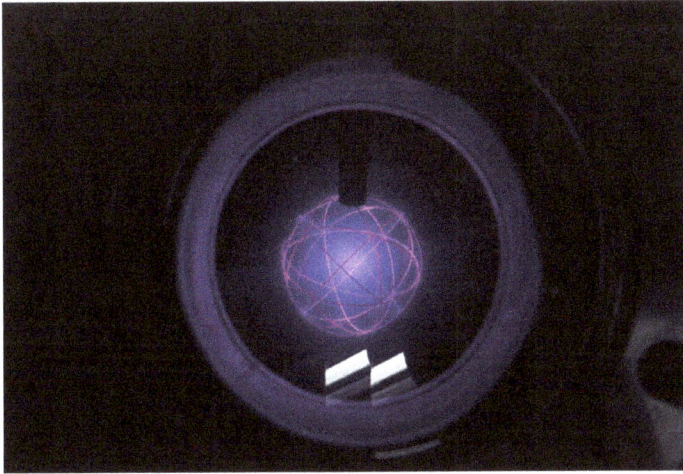

Electrostatically confined plasma (University of Wisconsin). The negative grid is visible through the viewing port.

reactions, IEC projects are still restricted to small scale experiments (Figure 9.9).

Electrodes in contact with the plasma are harmful. They contribute to particle losses and plasma cooling. This constitutes a major drawback that can be avoided by creating the potential well with electrons confined inside a cusped multipolar magnetic field (principle of R. W. Bussard's Polywell). Such configurations have other advantages. The radial electric field could be adjusted to take advantage of the resonances of the p–^{11}B reaction. For this and other aneutronic reactions, the design of the magnetic field geometry could be adapted for direct energy conversion. The concept has been extended to spatial propulsion [14].

A completely different approach for fusion reactions beyond D–T is standard inertial confinement. A recent experiment was reported in Section 7.7: a moderate laser compression of a spherical D–^{3}He capsule was implemented in order to use 14.7 MeV protons as a diagnostic: proton radiography of a hohlraum [15]. To reach ignition, a higher compression than in the case of D–T is required.

If confirmed, the radiative temperature of two billion degrees obtained with the Z machine would help. The ignition of the D–^{3}He and P–^{11}B reactions would be foreseeable. Although the energy released is 40%

smaller, there is some advantage in considering the latter reaction. The elements taking part in the reaction are not scarce resources. This is obvious for hydrogen. The average boron concentration in the Earth's crust is 10 ppm. Borates deposits have been found, the more so in Turkey (2/3 of the known reserves). Isotope 11 is the most abundant (80%).

Provided a higher compression is achievable, inertial confinement looks more promising than magnetic confinement in the perspective of fusion with fuels other than D–T.

Thermonuclear fusion has stimulated the imagination of scientists for more than half a century. Proving it to be feasible is still a long way to go even along the main trails: tokamaks ands laser-driven inertial confinement. There is ample time and possibly a real need to think out of the box and explore exotic ways.

References

1. European Physical Society position paper: *Energy for the future, the nuclear option.* http://www.eps.org/highlights/energy-for-the-future/.
2. C. Rubbia, *AIP Conf. Proceedings* No. 346 (1995); H. Nifenecker, O. Meplan and S. David, *Accelerator Driven Subcritical Reactors (ADSR)*, 2002, and references therein.
3. A. Sakharov, *Memoirs* (in Russian 1978), Translation R. Lourie, Alfred A. Knopf (1990).
4. L. M. Lidsky, Fission-fusion systems: hybrid, symbiotic and augean, *Nuclear Fusion* **15** (1975) 151–173; H. Bethe, *Physics Today*, May 1979.
5. P.-H. Rebut, in *La fusion nucléaire: de la recherche fondamentale à la production d'énergie?* Rapport de l'Académie des Sciences EDP-Sciences (2007).
6. M. Kotschenreuther, P. M. Valanju, S. M. Mahajan and E. A. Schneider, *Fusion Engineering and Design* **84** (2009) 83.
7. N. G. Basov, *Quantum Electron.* **23** (1993) 282.
8. https://lasers.llnl.gov/about/missions/energy_for_the_future/life/.
9. Jeffrey P. Freidberg and Andrew C. Kadak, *Nature Physics* **5** (2009) 370. Toni Feder, Physics Today **62**(7) (2009) 24.
10. For a complete story see: John R. Huizenga, *Cold Fusion, the Scientific Fiasco of the Century*, Oxford (1992).
11. B. Brunelli and G. Leotta, *Muon Catalyzed Fusion and Fusion with Polarized Nuclei*, Plenum (1987).
12. W. M. Nevins and R. Swain, *Nuclear Fusion* **40** (2000) 865.
13. R. P. Ashley *et al.*, *Fusion Science and Technology* **44** (2003) 564.
14. R. W. Bussard, 57th International Astronautical Congress (IAC 2006).
15. C. K. Li *et al.*, *Science* **327** (2010) 1231.

10

The Fusion Reactor

1. Achilles and the tortoise, a paradox revisited

Since the 1950s, the ultimate goal of controlled fusion research is an economically profitable power plant. Impressive progresses were made in magnetic (tokamaks) and in inertial (laser-driven) confinement. However, break-even is postponed to a remote future. Science and technologies are not fully mastered yet. On the contrary, fission power plants (Achilles) were implemented within a remarkably short time lapse in the 1960s and 1970s. Nowadays the development of nuclear technologies is much slower, specially in the case of the breeder reactors. Paradoxically, its constant slow pace could, by 2050, put fusion (the tortoise) abreast with fission power plants of generation 4.

At the end of 1938, fission was an unexpected discovery. Then, advances were quickly made: the first controlled chain reaction was obtained on December 2, 1942. As soon as 1954, nuclear reactors were delivering electrical power to the grid [1]. In the 1970s, nuclear power plants had taken a significant part in electricity generation. From the proof of principle experiment to GW power plants, less than 30 years were needed.

The birth of nuclear energy came among unusual circumstances. Once the effectiveness of a chain reaction was demonstrated, founding principles were firmly established thanks to outstanding scientists like Enrico Fermi.

Strong incentives stemmed first from World War 2, then from the so called "cold war". Rapid progresses were made possible by sizable budgets and the recruitment of talented people who created a new branch of engineering science. The corresponding industry developed in a hurry. Nuclear reactors of all types were conceived and tested first in secrecy within newly founded national laboratories. In the early 1950s, a massive declassification took place culminating in the Geneva conferences "Atoms for peace" (1955 and 1958). Textbooks were published. Private companies, du Pont de Nemours, General Electric, Westinghouse, etc. who contributed anonymously to the Manhattan project benefited from technology transfers in order to develop commercial reactors.

In the 1970s, the oil crisis brought further momentum to nuclear energy. In France, a vast nuclear power plants program was decided as a response to the first oil shock of 1973. By the turn of the century 80% of electric power produced in the country was of nuclear origin.

The growth of fission-driven nuclear energy was indeed very fast. Although drastic safety measures had to be taken, implementation of power plants was quickly considered as routine engineering. It was rapidly and falsely inferred from this piece of history that abundant subsidies and manpower are the key for the solution of technological problems. Actually this is only true when the underlying science is firmly established as it was the case for fission.

Nuclear fusion is a counter example. Enthusiastic pioneers did not realize how long the way towards a power plant would be. Indeed in the 1950s, plasma physics was still in infancy. Many fundamental problems were discovered that had to be solved. Thanks to dedicated fusion researches, significant advances were made in plasma physics, nowadays a recognised important branch of physical sciences.

Most fusion researches are carried on in national laboratories or within international collaborations. As it was seen in Chapter 5 about tokamaks, the progresses were quite rapid. There is no evidence that more subsidies and manpower would have accelerated the pace. Regretfully, decisions about ITER can be rated as conservative. In 1998, the initial project, aimed at ignition, was considered too costly and rejected. At that time, the threat of a climatic change was hardly perceived by policy makers. Would the decision be different today?

2. Conditions for a power plant

In a D–T fusion power plant, most of the released energy goes to 14 MeV neutrons escaping the plasma. This kinetic energy is then converted into heat in a blanket some distance away. Similarly to natural (the Sun) thermonuclear reactor, the energy released by manmade devices (tokamaks, inertial targets), is transported out of the plasma to a converter (Table 10.1).

In a future reactor, a circulating coolant will be used for heat extraction from the blanket and transport via a heat exchanger to a thermal engine, all steps routinely operating in fission power plants (Figure 10.1).

Table 10.1. Energy from a stellar or a fusion plasma.

	Energy transported by	Converter	Extraction
Solar fusion	Photons	Solar thermal collectors Concentrating mirrors Solar panels (photovoltaic)	Heat (coolant) Electric current
D–T fusion	Neutrons	Blanket	Heat (coolant)

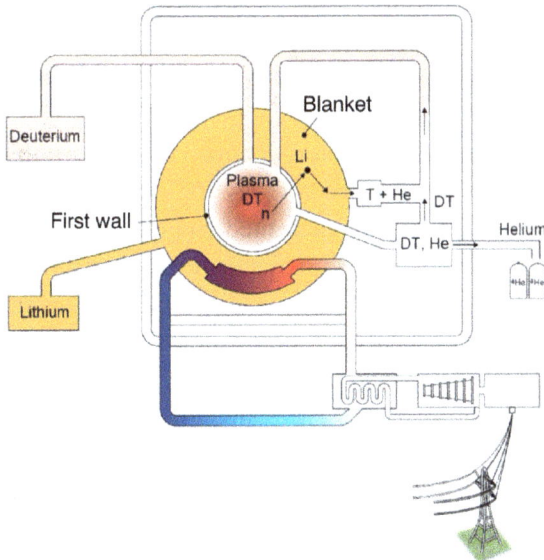

Figure 10.1. Basic scheme of a fusion reactor in a power plant.

The blanket has several functions:

- Transformation of the 14 MeV neutron flux into a heat flux to drive a thermal engine that powers an alternator
- Tritium regeneration (see below Section 10.7)
- Energy amplification in a hybrid system associated with other functions related to fissile fuel processing
- Protection of the outer world from neutron irradiation

The first wall facing the plasma should resist particle fluxes much higher than encountered in state-of-the-art nuclear technology. Neutrons and gamma rays (a small proportion) from the plasma are penetrating particles. The blanket thickness is determined by severe requirements about the protection of the environment. Activation by neutrons is unwanted outside the reactor. Altogether the future magnetically confined fusion reactor is to look like ITER: bigger and fitted with an electricity generator.

In the case of inertial confinement, fusion energy is released in a very short time: less than a nanosecond. Neutrons and debris originate from a point source. The blanket receives an almost instantaneous tremendously high energy flux. A liquid blanket appears as the most promising technology. The advantages are two fold: easy heat transport via the direct flow to a heat exchanger, elimination of problems with material damages. A popular concept is the lithium fountain, a hollow cylindrical cascade surrounding the exploding fusion capsule. The melting point of lithium is conveniently low (179°C) and tritium regeneration is obtained at least cost thanks to neutron capture. The thermonuclear reactor "Hylife", a proposal from Livermore along these principles [2], is represented on Figure 10.2.

For a fusion reactor in any confinement mode, gains and losses along the energy cycle are to be evaluated. Let Q be the fusion gain of the reacting plasma. Since 14 MeV neutrons enter exothermal nuclear reactions when colliding nuclei in materials, Q can be increased by the blanket, slightly in the case of ^{6}Li for instance. The blanket multiplying factor g is usually at most 1.2 except if fissile elements are incorporated (see Chapter 9). The main losses are due to the comparatively poor efficiency of a thermal engine, the second law of thermodynamics setting unavoidable limits. Let η_T be the overall efficiency ot the transformation of fusion power into electrical power. Now, a fraction f of the electrical power goes either to

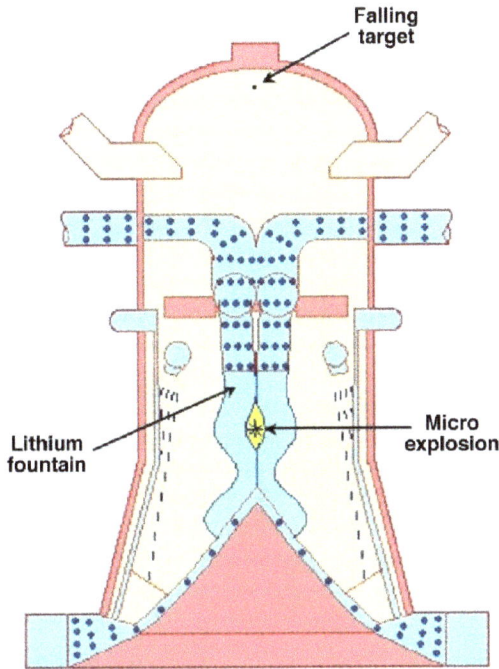

Figure 10.2. The "Hylife" thermonuclear reactor (after a LLNL document). Targets are released at a given repetition rate and fall freely along the axis of the lithium fountain. Arriving at the interaction point they are irradiated by laser or heavy ion beams. The liquid metal flow carries heat away.

plasma confinement and heating in magnetic confinement devices or to the driver for inertial confinement. In the latter case, the efficiency η_D of the entire compression process, i.e. the ratio between the internal energy of the imploded fusion fuel and the energy extracted form the power grid, is to be accounted for. Expected values of gains and efficiencies are reported in Table 10.2.

For a magnetically confined fusion reactor in steady operation, the products of all gains and efficiencies is unity, viz.:

$$Qf_N g\eta_T f = 1.$$

The system could be economically profitable if f were lower than 10%. Hence the condition:

$$Qg\eta_T > 10.$$

Table 10.2. Compared performances of magnetic and inertial fusion reactors.

			Reactor	
			Magnetic	Inertial
	Fusion gain	Q	40	150
	Gain in the blanket	g	1 to 1.2	1 to 1.2
	Efficiency of power generation	η_T	0.4	0.4
	Energy recycled fraction	f	<0.1	<0.1
Inertial	Repetition rate (s^{-1}),	ν		0.1 to 15
	Efficiency of the compression process	h_D		0.35

Table 10.3. Performances of hybrid fusion–fission reactors.

		Hybrid reactor	
		Magnetic	Inertial
Fusion gain	Q	2 to 5	30 to 60
Repetition rate $(s^{-1}$, inertial)	ν		0.1 to 15
Gain of the fissile blanket	g	10 to 80	10 to 150
Thermal efficiency	η_T	0.4	0.4
Energy recycled fraction	f	<0.1	<0.1
Efficiency of the driver (inertial)	η_D		0.05 to 0.35

Consequently Q is to be larger than 25.

By the same token, the profitability of inertial fusion could be obtained if the inequality:

$$Qg\eta_T\eta_D > 10,$$

were fulfillled, the ratios being expressed in energy values rather than power. The product $Q\,\eta_D$ is to be larger than 25 which implies a driver efficiency exceeding 40% for a gain of 100.

In both cases, a fissile blanket with a gain g between 10 and 20, acceptable values in routine operation, relaxes the requirements on the fusion gain. Values of significant parameters for hybrid magnetic or inertial reactors are listed in Table 10.3. A JET size tokamak would be convenient. Similarly, provided ignition experiments are successful, bigger than megajoule drivers would not be necessary.

As shown in Tables 9.1 and 9.2, future fusion reactors, either pure fusion or hybrids, will be gigantic devices as state-of-the-art fission reactors are. There is no such thing as "fusion in my basement". The giant scale implies huge investments and a lot of inertia including resistance to disturbing innovations.

The evolution of tokamaks towards bigger and bigger machines seems unavoidable. In the DEMO prototype reactor as designed in 2013, the plasma volume is 3 times the plasma volume in ITER. The expected electric power will be around 1 GW. If proven promising, the evolution of other magnetic configurations, compact tori, stellarators, is likely to be similar.

Facilities built for inertial fusion are already gigantic. Each one of the laser amplification hall of LMJ is longer than 100 m. Heavy ion acceleration will require kilometres of linacs and storage rings. High-power lasers and Z machines are pulsed devices whose electrical energy is stored in capacitors banks. Such a storage together with transmission lines needs a large volume. Capacitors are a poor storage system requiring 1 litre per KJ to be compared to 5 cm^3 of a storage battery or the 25 mm^3 of a liquid chemical fuel.

3. Energy cycles

A simplified energy cycle was imagined by J. D. Lawson in order to set up his celebrated criterion (see Chapter 2). After ignition, the fusion reaction is self-sustained and a steady regime holds. Then cycles for fusion and conventional power plant look identical and can be evaluated using power rather than energy (Figure 10.3).

For a tokamak, part of the electrical power is not delivered to the grid but recycled. Indeed, magnetic field generation, current drive and auxiliary equipments need power, altogether a fraction f of the produced electrical power. The energy delivered to the grid is proportional to the time the plant is operating in steady regime. This time is to exceed the minimum defined by Lawson criterion.

In ITER and future power producing tokamaks as well, no ignition is planned. The performance is given by the fusion gain Q, ratio of the fusion power to the power of auxiliary heating devices. In ITER Q will be 10.

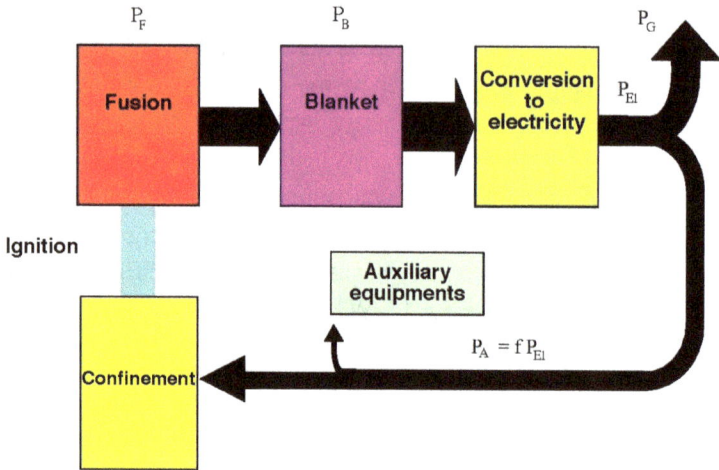

Figure 10.3. Energy cycle of a magnetically confined fusion reactor. After ignition the fusion reaction is self sustained. A steady regime holds in which a fusion power P_F is absorbed by the blanket. The heat power P_B estracted from the blanket is converted into electric power P_{EL} a fraction f of which is recycled for confinement and auxiliary equipments. The power sent to the grid is $P_G = P_{EL}$ (1-f).

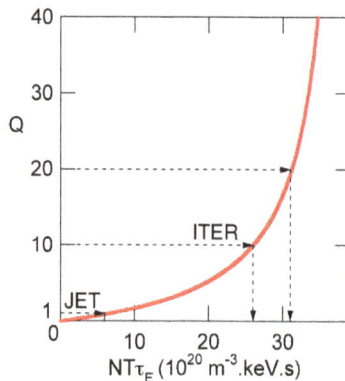

Figure 10.4. Fusion gain Q of a tokamak versus the triple product (semi-empirical plot, τ_E is the energy confinement time).

It was slightly below unity in JET. Actually, Q is a function of the triple product $nT\tau$ as shown by the semi-empirical plot of Figure 10.4.

Ignition corresponds to the divergence of Q at some 40×10^{20} m^{-3} keV s. At lower values of the triple product, Q can be high enough for a

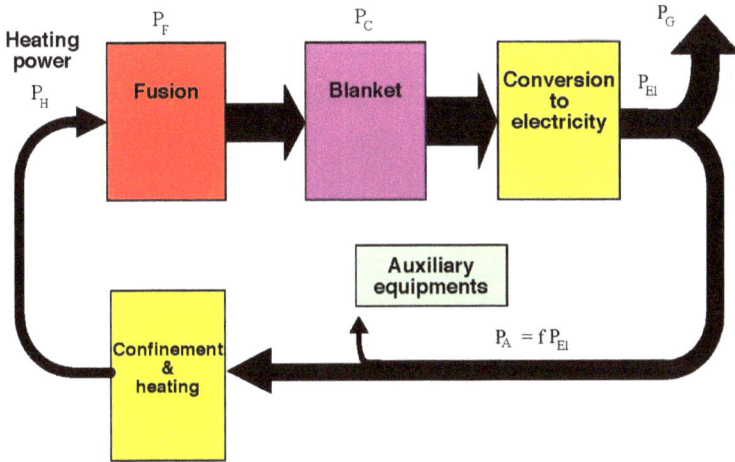

Figure 10.5. Energy cycle for a magnetically confined fusion reactor without ignition. The recycled power f P_{EL} serves also at plasma heating, a major difference with the case of Figure 10.3, ignited plasma.

closed cycle producing net power according to the scheme of Figure 10.5. Note the Lawson criterion does not imply ignition. Due to the necessary auxiliary heating, a self-sustained thermonuclear reaction is no longer needed in future tokamaks. Although designed for a power plant, DEMO, which could be the next step after ITER does not include ignition.

On the contrary, ignition is mandatory in the case of inertial fusion. Then the fusion power is the product of the released energy per shot times the repetition rate ν (number of shots per second). The fusion gain Q is the ratio between the fusion energy and the internal energy brought to the fuel until the micro-explosion occurs. The recycled energy goes mainly to the driver and to the fast ignitor if any according to the cycle represented on Figure 10.6.

The way Q depends upon the laser energy used for compression is known after numerical simulations. As shown in Figure 10.7, the gain is very sensitive to the modus operandi: direct or indirect drive, implosion velocity, central or fast ignition.

The figure evidences the advantages of fast ignition: minimization of the laser energy necessary for compression, better conditions for ignition, high target gain. However the 100 kJ required by the fast ignitor are to be included in the overall energy balance. In the European project HIPER, the

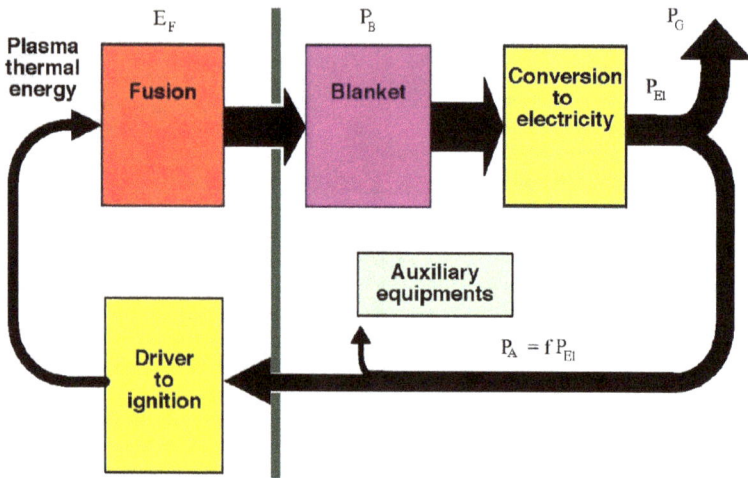

Figure 10.6. Energy cycle for inertial confinement fusion. Fusion energy E_F is released at each shot. The fusion power P_F is the product of E_F times the repetition rate ν. Every explosion requires compression energy from the main driver plus energy for fast ignition if any. The vertical lines separates shot by shot (left) from continuous operation (right).

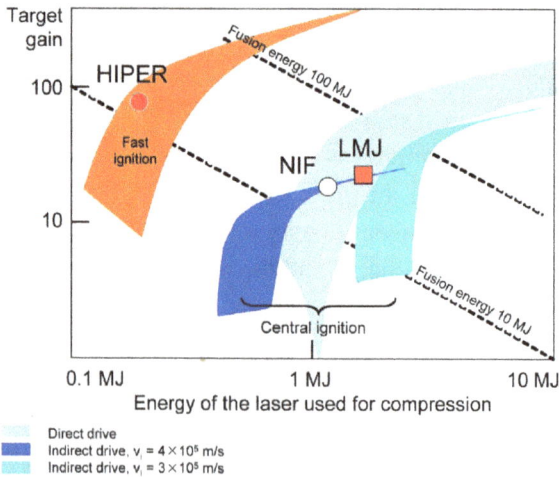

Figure 10.7. Target gain versus the energy of the laser used for compression. Coloured areas correspond to results of innumerable computer simulations. Several cases are considered: compression followed by fast ignition, compression leading to central ignition after direct or indirect drive (2 different implosion velocity v_i). Expected performances of megajoule lasers, NIF and LMJ, and project HIPER are given.

laser energy for compression is 200 kJ only. In order to develop the very high power short pulses for ignition, the LIL laser is to be refurbished; the PETAL (PETawatt Aquitaine Laser) aims at 3.5 kJ, 0.5 to 5 ps pulses [3].

The best case implying central ignition is obtained with direct drive. A laser energy higher than 3 MJ would lead to a fusion gain Q close to 100.

Driver performance, high repetition rate target injection, D–T spherical compression up to ignition conditions are key issues to be mastered before an inertial fusion reactor is at hand. Three families of drivers have been considered: lasers, heavy ion beams, and Z-pinches. Which one is the more likely to lead to competitive reactors is still an open question.

Heavy ions and Z-pinches are solutions for the future. So far investments in heavy ion fusion are staying well below investments in megajoule lasers. Research is still at a preliminary stage: basic physics experiments, drafts. On the contrary, the technology of Z-pinches is already effective in the megajoule range. In both cases, no decision has been taken about developments of inertial fusion reactors.

Many years ago, high-power lasers were chosen as the driver for full-sized experiments on inertial fusion. Laser-driven experiments are likely to be used for progress assessments along the road to be followed during the decades to come towards a reactor.

4. Fusion in the nuclear fuel cycle [4]

According to the present state of knowledge, fusion–fission hybrid reactors could reach the industrial stage much earlier than pure fusion. As far as energy is concerned, the difference lies in the high gain of the blanket. Fitted with a fissile blanket, a tokamak the size of ITER or a target imploding with a laser of the megajoule class with a suited repetition rate would be the inner core of a reactor whose electric power would compare with an EPR (European Pressurized Reactor). However, the investment would be higher. The profitability of the project is then questionable. It could be enhanced, with a design including the production of fissile isotopes and the incineration of unwanted elements.

Fusion–fission hybrids would thus play the same role as ADS (Accelerator Driven Systems) in the nuclear fuel cycle. Their main part would be the elimination of highly radioactive components of spent nuclear fuels: fission products, minor actinides, plutonium (including weapon grade), whilst extracting energy from the blanket. The fusion–fission hybrid could contribute to the solution of both problems: nuclear waste and proliferation.

In the USA, there is a renewed interest in hybrids in the more general context of a possible "nuclear renaissance" [5]. Indeed, impending threats of oil shortage and climate change favour energy sources that do not emit carbon dioxide. In the US the nuclear fuel cycle is of the once through type: no reprocessing and plutonium is among radioactive waste that are to be stored. Besides radioactivity, proliferation is thus made easier.

Proposals about hybrids integrate fusion in the general frame of nuclear technologies. It can be used for power production and incineration of nuclear waste. Since part of the incoming neutrons are thermalized, fissile elements could be created inside the blanket. The three functions of hybrids, power generation, breeding of fissile elements and radioactive waste elimination, can be represented on a chart (Figure 10.8) L. Lidsky proposed in the 1970s [6].

Preliminary studies lead to the concept of an industrial nuclear complex comprising a fusion reactor, several pressurized water fission reactors (PWR) of the third generation, a reprocessing plant and in some instances a fast neutron fission reactor of modest power. All the facilities being built on a single site, no external transport of radioactive materials is needed. For instance, a complex devoted to the thorium cycle would be organized around a hybrid fusion–fission reactor whose fusion part is an ITER size

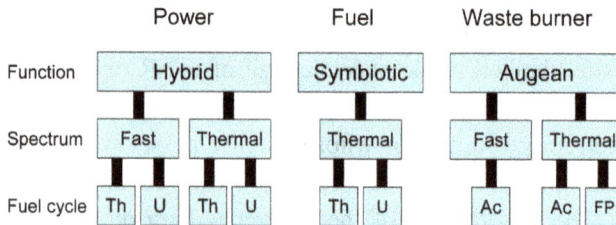

Figure 10.8. Lidsky's chart: the three functions of fusion–fission hybrids. Part of the 14 MeV neutrons are thermalized in order to ensure complete transmutations to elements that will enter in thorium and uranium fuel cycles. Simultaneously, other actinides (Ac) and fission products (FP) are eliminated.

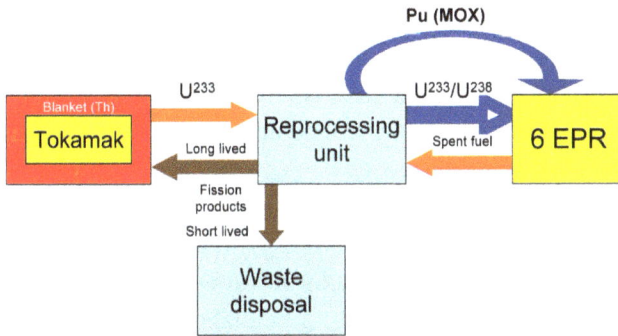

Figure 10.9. Basic scheme of a nuclear complex implementing the thorium cycle. Thorium is converted into Uranium-233 in the blanket of a tokamak hybrid facility. Long life fission products can also be transmuted in the blanket. In the multi purpose reprocessing unit, plutonium from the EPR is also converted into fuel (MOX). No radioactive materials are transported outside.

tokamak: 500 MW thermal [7]. The blanket would incorporate thorium and lithium. 20 kg of tritium and 5.5 tonnes of Uranium-233 would be produced every year. Once separated and mixed with Uranium-238, Uranium-233 would enter active elements suited to light water fission reactors. The annual production of fissile isotopes would be sufficient to power 6 EPR (Figure 10.9).

Plutonium produced in the EPRs can also be reprocessed into new nuclear fuel (MOX) and is thus eliminated. Short-lived radioactive by-products are stored onsite. Long-life products are eliminated in the blanket of the hybrid reactor.

Inertial confinement can also take part in the nuclear fuel cycles. LIFE or InZinerator look suited to the elimination of nuclear waste. As it was seen in Chapter 9, LIFE is a once through device, the 30 years full power production phase being followed by a 10 years incineration phase. Only two functions among three will be implemented in such devices.

5. A glimpse of the future

The immediate future of fusion projects is planned. ITER is under constructions but milestones have shifted. Initially, the first plasma was due in 2015 followed by operation with tritium by 2020. Now the pace of this complicated international endeavour is slower than expected. After the latest meetings of the steering board, the new schedule is: first plasma in 2020, tritium by 2026. Ample time is left for preassembly and preliminary tests of major components of the machine: vacuum vessel, magnetic field generation, modular blanket . . .

Beyond ITER, DEMO is to demonstrate a gain higher than 30 together with power generation. It would correspond to the first power generation to the grid from a fission reactor achieved at Obninsk (Russia, then USSR) in 1953. The design of DEMO is to be elaborated whilst ITER is being constructed and experiments begin. ITER 1998 could serve as a baseline. A possible timeline for ITER and further tokamaks [8] is presented on Figure 10.10.

If it were a tokamak, DEMO would look like ITER, bigger with a whole set of circulating fluids and heat exchangers. Design, construction and operation of the facility will need time that could be augmented by the delay in decision making. DEMO could be operational after 2035. With a current exceeding 20 MA, it could generate 1.5 GW of fusion power, enough for delivering some 500 MW electrical power to the grid, notwithstanding the lack of profitability.

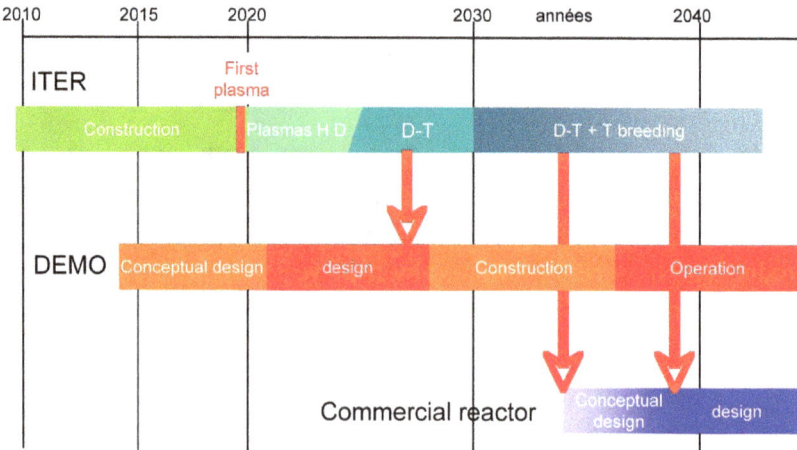

Figure 10.10. Tokamak roadmap towards a commercial reactor. The final design of DEMO is to benefit from the early experiments on ITER. Routine operation of ITER, then of DEMO, are to be accounted for in the design of a commercial reactor.

After DEMO, the development of industrial reactors will be a rather slow process: 20 years are usually needed. In the second half of the 21st century, the fusion reactor should prove commercially viable.

Will the reactor be a tokamak? The question is open.

For the last 40 years, research on magnetic fusion concentrated on tokamaks. The reason is the device had proven better than any other configuration in the late 1960s. Will it be still true a century later? In ITER, energy extraction and tritium breeding will be tested. Hopefully, reliability and long time behaviour will be examined.

Meanwhile, other magnetic configurations could mature over time. For instance, the stellarator has its own complexities: the magnetic field is to be calculated and implemented with a challenging high precision. The resulting twisted shape of coils and blanket is hard to manufacture. Furthermore, the high aspect ratio increases the overall diameter of the plasma ring at given plasma volume. However, advances in numerical computations were recently made [9]. If Wendelstein 7-X proves successful, the stellarator would appear as a good contender for DEMO (Figure 10.11) with an advantage. Indeed, continuous operation is possible contrary to the tokamak, a pulsed machine which needs sophisticated items such as a high power pulse generator with prescribed pulse shaping, a transformer and current driving devices.

On the inertial fusion side, NIF has been running at full power but ignition was not achieved during the 2012 National Ignition Campaign. No blanket implementation is planned at NIF or at LMJ, contrary to ITER

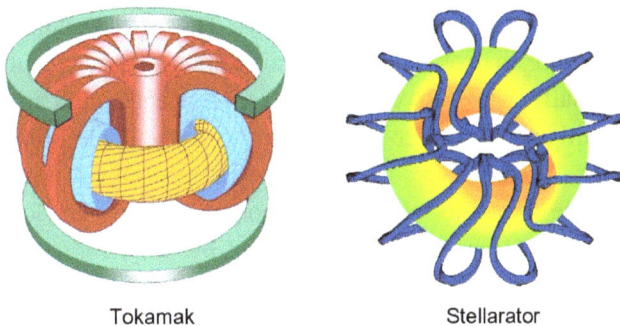

Tokamak   Stellarator

Figure 10.11. Two contenders for DEMO, a full-size fusion reactor with 2000 m^3 of magnetically confined plasma.

in which the neutron absorbing blanket is aimed at reactor studies. The goal of NIF or LMJ is the demonstration of ignition through a small number of full-size shots. Given the state of the art, experiments on the transfer of fusion energy would look premature.

However, studies have been carried on for a long time on conceptual inertial fusion reactors [10]. The blanket is mandatory. Small scale experiments are running on specific points such as liquid lithium flows. Other researches deal with a lot of key issues for a reactor. Which target design is best suited to obtain expected gains in the range $30 < Q < 200$? Which process of mass production will deliver costless targets? Which driver will be closer to reactor requirements: efficiency and repetition rate? Which blanket design will be the most efficient for energy extraction? Among others, HIPER is a project intended at answering such questions.

6. A touch of economy

Fusion is a potential energy source eventually aiming at delivering power to the grid. The cost of the energy unit is to be consistent with the cost of energy from other sources. Reliable information about the economics of magnetic fusion will be available only after running DEMO for some time, i.e. after 2050. The status of the energy market in a remote future is unknown. However, data are available and trends appear which allow one to cautiously put forward some predictions [11].

Nowadays, the energy consumption of mankind comes mostly (about 80%) from fossil fuels: oil, coal and gas. This is unsustainable. A first reason is a severe environmental constraint: emissions of greenhouse gases (GHG), specially carbon dioxide (CO_2) from coal and hydrocarbon burning, are to be mitigated in order to avoid an uncontrollable climate change. Furthermore, fossil fuel resources in the Earth crust are finite. There is some evidence that peaks in hydrocarbon production are looming [12]. A maximum in oil production is expected by 2020 followed by a slow decay till the end of the 21st century. A similar evolution, shifted by a few decades, is to occur with gas and possibly with coal, the most abundant, the cheapest but the worst GHG emitter fossil fuel. Unconventional resources, shale oil and shale gas, just postpone the peaks. The future of coal depends upon implementation of expensive Carbon Capture and Storage (CCS) technologies. Very likely, it will be more and more difficult for a potential supply to match an increasing demand due to an increasing population a larger fraction of which is developing. Consequently, energy prices are to be pushed upwards in the decades to come with possible disastrous effects on the economy. The higher the cost of fossil fuels, the more competitive will be other energy sources.

The cost of the KWh is made of several parts: investments costs including savings for future decommissioning and dismantling, running costs including fuels and reprocessing for nuclear power plants that are not once through. In fossil fuel power plants, the investments are comparatively low unless CCS is implemented. So, ignoring CCS, the cost of the KWh is very sensitive to the price of oil, gas and/or coal. On the contrary, in generation 3 (EPR for European Pressurized Reactor) or 4 (fast breeders) nuclear power plants, investments are important, the more so since the power of a single reactor is high, typically 3 to 4 GW thermal. Accordingly, the cost of the KWh is loosely dependant upon the price of uranium.

As far as the economy is involved, fusion will not be different from other nuclear technologies. Investments are to be the main component in the cost of the KWh. Other predictions are obviously premature. However economic studies were carried on. The results are to be taken cautiously [13]. An estimate of the investment per rated KW is $ 4000, about the same as for the first generation 3 fission reactors. Other estimates deal with the cost by 2030–2050 of 1 KWh delivered by 1 GW electrical-power plants powered using different energy sources: magnetic confinement (tokamak) fusion, inertial confinement fusion with various drivers, PWR fission reactors of EPR type, coal with and without CCS, and natural gas (Figure 10.12).

How optimistic such predictions are is difficult to tell. Criteria for fusion reactor were met: gain exceeding 40 or ignition for tokamaks, global efficiency of the driver and fuel compression exceeding 30% for inertial confinement. Effects of the complexity of fusion devices were presumably underestimated.

Hybrids could improve the balance. True, incorporating fissile materials increases the cost of the blanket. Hopefully, this is to be overcompensated by the gain and savings due to lesser constraints on the first wall. Altogether, the investment per rated KW would be reduced with respect to a pure

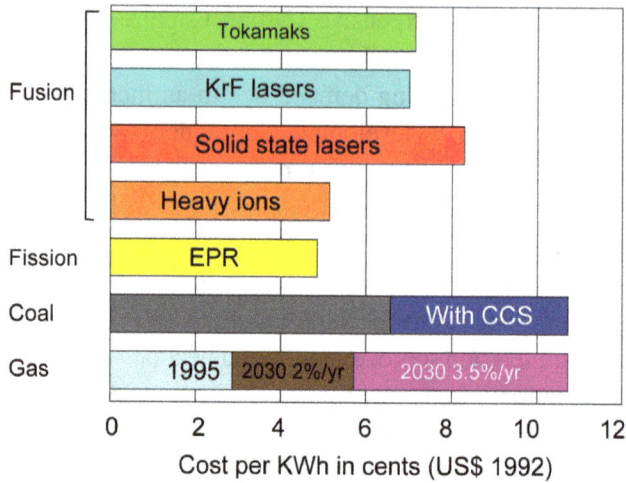

Figure 10.12. Estimates after reference [13], of possible costs of the KWh by 2030, from 1 GW plants powered by magnetically or inertially confined thermonuclear reactors. Comparison is made with other sources of primary energy: coal with or without CCS, gas with or without annual percentage of price increase (estimates were made long before shale gases entered the scene in the USA).

fusion reactor. Hence the competitive candidates for future nuclear energy production are: generation 4 fission reactors, ADS, fusion–fission hybrids, pure fusion reactors.

7. Tritium breeding and resources

In the near future, all fusion reactors will use the D–T reaction. Deuterium is abundant in nature. On the contrary, tritium is exceedingly scarce. The nucleus is radioactive: β^- decay with half-life 12 years producing a ^{3}He, an electron and an antineutrino. Hence, only some traces have been found in the upper atmosphere. A fusion reactor delivering 1 GW (7–8 TWh/yr) to the grid would need every year only 100 kg of deuterium and 300 kg of lithium, the resource for tritium production.

Tritium is obtained after neutron impact on lithium (see Chapter 2), usually in dedicated fission reactors. An alternative solution is tritium breeding in a lithium loaded blanket thanks to the 14 MeV neutrons emitted by the D–T reaction. Beryllium, a neutron multiplier is added in order to control neutron fluxes. In ITER, lithium-doped ceramics will be incorporated in the blanket. An alternative solution is the eutectic lead–lithium. Since the melting point is at 235°C, it can be considered only for power-producing reactors, beyond ITER. The tritium breeding is enhanced using lithium with a proportion of isotope 6 versus isotope 7 higher than in nature. Then more than a tritium nucleus per 14 MeV neutron can be obtained, thus reducing the cost of tritium which should be accounted for in the general economy of D–T fusion.

Proven reserves of lithium [14] are 28 million tonnes of metal whilst the demand (mainly for Li-ion batteries) is 16000 tonnes per year. Such reserves would ensure for 3000 years the electrical power production of the entire planet at the 1995 level [15]. In 2010, Bolivia was the main supplier. However, the resource is poorly known. For instance there is some evidence for reserves in Afganistan comparable with Bolivia. Lithium is also present in ocean waters. The concentration is low, 0.17 g/m^3 for a total resource of 233 billion tonnes. Extraction would be more costly compared to deuterium.

In the Sun, deuterium created by the proton–proton reaction, disappears immediately in further reactions. None is found in the solar wind. Therefore, deuterium on earth cannot be of solar origin. Actually, it was created in the early universe when, shortly after the "big bang", a synthesis of light elements occurred. The measured abundance of deuterium is used in the determination of the age of the Universe. Due to its primeval origin, deuterium is the least renewable of all fuels. Now, seawater contains 200 ppm of heavy water, i.e. a (large) glass per cubic meter. The total amount of deuterium in the oceans is 45000 billion tonnes, enough for billions of years should fusion be the unique energy source to power mankind.

Similarly, the fuel problem of the D–^{3}He reaction is the availability of ^{3}He. Only minute quantities can be found in the Earth crust. Since the Moon has no atmosphere, its surface is constantly bombarded by solar wind particles, hydrogen and helium, that have accumulated for 4.5 billions of years in the first layers of rocks. Although a minor component of the particle flux, the amount of Helium-3 economically accessible on the moon is estimated at around a million tonnes, enough for powering fusion reactors during millenaries. Notwithstanding the complexity and the cost of the undertaking, gathering of ^{3}He was considered a valuable objective for resumed manned missions to our satellite [16]. Within the solar system, the atmospheres of giant planets are other reservoirs of ^{3}He that could be a resource in a remote future.

Among all energy sources using fuels, the resources for nuclear fusion are by far the most abundant. Fusion will be a solution for the future of power production provided a reactor proves economically competitive when arriving on the market.

References

1. See e.g.: http://www.euronuclear.org/info/npp-ww.htm.
2. R. W. Moir, *Particle Accelerators* **37–38** (1992) 467.
3. http://www.hiper-laser.org/ et http://petal.aquitaine.fr/.
4. D. G. Cacuci, (Ed.), *Handbook of Nuclear Engineering, Vol. 5 Fuel Cycles, Decommissioning, Waste Disposal and Safeguards*, Springer (2010); an updated account of nuclear fuel processing can be found in: *Le nucléaire expliqué par des physiciens*, B. Bonin, ed., EDP-Sciences (2012), in French, to be published in English translation by World Scientific Publishing Co.
5. J. P. Freidberg and A. C. Kodak, *Nature Physics* **5** (2006) 370; T. Feder, *Physics Today*, **62** (2009) 24.
6. L. M. Lidsky, Fission-fusion systems: hybrid, symbiotic and augean, *Nuclear Fusion* **15** (1975) 151–173.
7. Adapted from W. M. Mannheimer, Can Fusion and Fission Breeding Help Civilization Survive?, *J. Fusion Energy* (2006).
8. K. Ikeda, *Nucl. Fusion* **50** (2010) 014002.
9. P. R. Garabedian and G. B. McFadden, *J. Res. Natl. Inst. Stand. Technol.* **114** (2009) 229–236.
10. M. J. Lubin and A. P. Fraas, Fusion by laser, *Scientific American*, June 1971.
11. See e.g. B. Richter, *Beyond Smoke and Mirrors: Climate Change and Energy in the 21st Century*, Cambridge (2010); J. L. Bobin, H. Nifenecker, C. Stephan, *L'énergie dans le monde, bilan et perspectives*, EDP-Sciences, 2nd edition 2007, in French.
12. M. K. Hubbert, Nuclear Energy and the Fossil Fuels, *American Petroleum Institute Drilling and Production Practice*, Proceedings of Spring meeting, San Antonio, 1956; see also Kenneth J. Deffeyes, *Hubbert's Peak: The Impending World Oil Shortage*, Princeton University Press (2002).
13. UCRL-MI-125743 (1997), download at http://fusioned.gat.com/education_notebook/images/pdf/ife.pdf.
14. R. K. Evans, An abundance of Lithium, March 2008, www.worldlithium.com.
15. J. P. H. E. Ongena and G. Van Oost, *Transactions of Fusion Technology* (1998).
16. H. H. Schmitt, *Return to the Moon*, Copernicus books (2006).

Epilogue

How long before a nuclear fusion reactor will be available? Researches started in the early 1950s. More than a half century later, the layperson might think no effective result was obtained, only pharaonic facilities being built whose objectives are dubious: ITER and megajoule lasers. In the 20th century, fusion used to be advertised as an energy source to be operational in the next 10 to 20 years. Nowadays predictions about an industrial development, if any, state it could take place in the second half of the 21st century.

Before that, evidence should be obtained that fusion is technically and commercially viable. In 2013, no one can be sure it will happen. "Moments of truth" are still to come both for magnetic and inertial confinements.

From present data and trends, scenarios can be outlined. The most predictable are with tokamaks. A large international community is at work with a road map that is planned for the next decades. The moment of truth is to come by 2026: will a fusion gain Q of 10 be achieved during 400 s in reliable and reproducible shots? The next milestone, a successful DEMO is not expected before 2040.

Is tokamak the best choice? According to the progresses made from 1969 to 1997 (Figure 5.8), the answer would be yes. Indeed during this period the triple product increased faster than the number of transistors on a computer chip. The tokamak appeared even more promising than it was in 1968. Afterwards came an era of stagnation. Designing and building a gigantic international high tech facility are lengthy tasks. The decision making processes in an international organization are also very slow. As time elapsed doubts aroused.

Eventually, other devices could prove more suited to an industrial implementation. Should hybrids be seriously considered, compact tori, for example, could be a convenient basis, producing energy and incinerating radioactive waste within the general frame of fourth-generation nuclear energy.

There is no such thing as an international project in the field of inertial fusion. Research with megajoule lasers, partly classified, is conducted within national laboratories. The most advanced on the fusion trail is NIF at Livermore (USA) with a completed and running laser. Was the National Ignition Campaign of 2012 the moment of truth? Although ignition has not been achieved yet, there is hope that further experiments might be successful. An extension of the field of investigation would help: direct drive, ignition schemes other than central hot spot, enhancing the energy delivered by megajoule lasers, the Z machine as a driver. . .

Missing ignition in the near future would imply building more powerful drivers. Very likely, the evolution of high-power glass lasers could come to an end as far as a fusion-energy source is involved. Nevertheless, there would be a lot of physics to be done with such facilities.

Should ignition be realized, the next step would be high efficiency (>30%), high repetition rate (>1 Hz) drivers. Fusion–fission hybrids could be the fast-track solution towards industrial devices. Other requirements are low cost targets and drivers fit to the job. Still a long way to go!

Tokamaks and laser-induced inertial confinement have been the main approaches towards controlled fusion receiving large funding and manpower. However, room is left for innovative projects if any that might open shortcuts.

The energy future of mankind is rather bleak. Nowadays, fossil fuels, oil, gas and coal, are the primary source, amounting to 80% of total. Since resources in the Earth's crust are finite, fossil fuels wont match energy demand for long. However, predictions about a coming peak oil production are not unanimously endorsed. Furthermore, combustion of fossils results in severe greenhouse gas (GHG) emissions. Since the early days of the first industrial revolution (late 18th century), the concentration of GHG in the atmosphere raised by some 40%. It is still increasing at the same rate. Consequently, a global warming of anthropic origin is threatening. The problem has been examined at the United Nations level through the United Nations Framework Convention on Climate Change (UNFCCC). On the other hand, the Intergovernmental Panel on Climate Change elaborate reports in which results scientists obtained all over the world are compiled. So far no significant action has been implemented.

There is no such thing as a "silver bullet", i.e. a single technology whose use would solve all energy problems expected to appear in the middle of this century, i.e. match the needs while strongly reducing GHG emissions. Unless "degrowth" policies are enforced, it will be mandatory to combine every available technology to match the required amounts at all relevant scales: region, country and the planet as a whole.

All substitutes to fossils have setbacks. Nuclear energy is almost GHG free. However, it is not accepted everywhere in the world. Fission power plants are running satisfactorily but for a few dramatic accidents. Furthermore radioactive wastes are perceived as a major threat. Provided it works, nuclear fusion could contribute to the solution of these problems. Pure fusion reduces drastically the amount of radioactive waste. Fusion–fission hybrids could contribute to the incineration of such unwanted elements. In all cases, no runaway is possible in a fusion device so no major nuclear accident is to be feared. Although controlled thermonuclear fusion is still a long way ahead, it might represent the ultimate clean energy source and is worth further research and developments.

Glossary

Ablation front   Surface separating unperturbed or shocked matter from the plasma flow driven by high-intensity radiation or by a particle beam. Similar to a rocket, shocked matter moves in the same direction as incoming radiation or particles.

Ablator   Layer made of light materials to be transformed into a plasma and set into motion under a high radiative intensity. Starting from an ablation front, a rarefaction wave is created in which plasma propagates in a direction opposite to the incoming light.

Adiabatic compression   Compression of a fluid without any heat exchange with the external world. A reversible adiabatic process is called isentropic.

Accelerator Driven System (ADS)   Sub-critical assembly of fissile elements irradiated by neutrons originating from a spallation source: 1 GeV protons impinging onto a lead target.

Aspect ratio (magnetic fusion)   Ratio of the major radius to the minor radius at the boundary of a toroidal plasma ring.

Blanket   Structure surrounding a reacting plasma aimed at: absorption of escaping reaction products and transfer of their energy to a coolant; tritium breeding; shielding. In a fusion–fission hybrid, the blanket includes fissile materials that contribute to energy amplification. It could also enter in the nuclear fuel cycle.

Break-even (magnetic fusion)   The point at which fusion power released in the plasma equals the input power of auxiliary heating.

Break-even (inertial fusion)   The point at which fusion energy released by the exploding pellet equals the energy invested in fuel compression and heating.

Chain reaction   In neutron-induced fission of a heavy nucleus, more than 2 neutrons are emitted. These secondary neutrons can induce more fissions.

Compound nucleus   Most unstable object resulting from the coalescence of the nuclei entering a nuclear reaction.

Critical density   A plasma is at critical (cut-off) density if the electron plasma frequency is equal to the frequency of incident electromagnetic waves. The waves propagate through the plasma only if the density is below critical. The critical density is also a resonance density.

Criticality   State of a fissile assembly in which gains in neutron number exactly compensate the losses. The effective multiplication coefficient is then unity. In a fission reactor, this state is to be both sustainable and stable with respect to any perturbation that might occur.

de Broglie wave   A massive microscopic object has a dual nature: both wavelike and particle like. The wavelength is related to the momentum by the celebrated formula: $\lambda = h/p$.

Debye length (or radius)   In an electrically neutral plasma, distance at which the Coulomb force induced by a point charge is $1/e$ the value in vacuo. At longer distances, the electrostatic field of a charge is strongly attenuated.

DEMO (DEMOnstration Power Plant)   Fusion reactor coupled with alternators, hopefully the next step after a successful ITER.

Dielectric constant of vacuum   Denoted by ε_0, a coefficient in the Coulomb force expressed in SI units.

Divertor Magnetic device using an X point in the meridian cross section of a special magnetic surface: the separatrix (Figure 4.6 b). Unwanted charged particles escape the confinement and are guided towards absorbing plates.

Edge localized modes Instabilities in the boundary layer of a tokamak that are detrimental to energy confinement.

Effective multiplication coefficient In an assembly of fissile elements where a chain reaction takes place, ratio between numbers of neutrons in generations n + 1 and n, taking into account all gains and losses. This ratio is unity in a steady regime.

Electron plasma frequency Frequency of spontaneous oscillations of the electron gas in a plasma. It is given by the formula (SI units):

$$\omega_p^2 = \frac{n_e e^2}{\varepsilon_0 m_e}$$

where n_e is the number of electrons (with charge e and mass m_e) per unit volume, ε_0 is the dielectric constant of vacuum.

Energy levels In an atom or a molecule, electron states have a precisely defined energy. In radiation absorption or emission, an electron jumps from an energy level to another one.

Ferritic/martensitic Apply to steel containing certain phases of iron carbon alloys: ferrite and martensite.

Fission Process in which a nucleus is split into two main fragments plus neutrons and radiation. Fission is generally induced by a neutron impact. Spontaneous fission is also possible. The number of emitted neutrons varies from 2 to 5 per fission of a heavy nucleus such as uranium. With respect to their mass number, fission products have a two-peak distribution with maxima around 90 and 140.

Hohlraum In indirect drive inertial fusion, cavity aimed at confining thermal radiation resulting from laser interaction with the walls or from any other pulsed power source.

Hot spot A tiny hot region within a compressed pellet that serves to ignite the main fuel. A hot spot might occur spontaneously at the centre of an imploded sphere. It can also be created by an external energy source.

Inertial Electrosatic Confinement (IEC) Confinement scheme in which, thanks to electrostatic fields, ions gather in the centre of sphere. Their energy can be adjusted to match resonances in the cross sections of fusion reactions.

In-flight aspect ratio (inertial fusion) Ratio between the distance of the centre and the thickness of an imploding shell.

Ion, ionized An atom or a molecule losing or capturing at least one electron is transformed into a positively or negatively charged ion. Plasmas are ionized gases with enough ions and electrons to be sensitive to electric and magnetic fields. A fully ionized plasma is deprived of neutrals. Nucleons are not necessarily stripped off of all their peripheral electrons.

Isotope Isotope nuclei correspond to one and the same chemical element. Given the number of protons, they differ by the number of neutrons.

International Thermonuclear Experimental Reactor (ITER) International fusion project launched in 1986 after a Gorbatchev Reagan summit. The Tokamak is being built at Cadarache (France).

Joint European Tokamak (JET) Third-generation Tokamak built at Culham (Great Britain) and running since 1983. Until 2000 (retrocession to UK), it was a European laboratory.

Kinetic energy Energy associated with motion. In classical mechanics, it equals half the product of the mass and the velocity squared.

Laser MégaJoule (LMJ) 240 beams laser being built in Centre d'Etudes Scientifiques et Techniques d'Aquitaine (CEA, France).

Magnetic permeability of vacuum Denoted by μ_0, a coefficient in the magnetic force between electric currents expressed in SI units. The product $\varepsilon_0\mu_0$ is equal to the inverse of the speed of light in vacuo squared.

Magnetic surface In a plasma immersed inside a magnetic field, a constant pressure surface that also contains field and current lines.

Mega Ampere Spherical Tokamak (MAST) Compact torus running at Culham, Great Britain, the same laboratory inside which JET is also running.

Magnetohydrodynamics (MHD)   Science of conducting fluids that are sensitive to magnetic fields.

Microballoon   Hollow glass sphere with a diameter smaller than 1 mm. This industrial product is obtained by blowing gas through liquid glass and solidification of the resulting bubbles. After calibration and control of the tightness, DT filling is obtained by high-temperature gaseous diffusion.

Murphy's law   "If any thing can go wrong it will". This sentence reflects the perceived perversity of the Universe.

National Ignition Facility (NIF)   196 beams laser with energy close to 2 MJ running at LLNL, California (USA).

National Spherical Torus (NST)   Compact torus running at Princeton Plasma Physics Laboratory (USA).

Nuclear proliferation   The spread of nuclear weapons, fissile material, and weapons-applicable nuclear technology and information to nations not recognized as "Nuclear Weapon States" by international treaties.

Planck's formula   Energy E of an electromagnetic radiation quantum is proportional to the frequency ν according to the formula: $E = h\nu$, where h is a very small universal constant. $h = 6.625 \times 10^{-34}$ in SI units.

Potential barrier   Image of the repulsion between electrically charged microscopic objects with same signs.

Radioactive waste   By-products of nuclear power generation and other applications of nuclear fission or nuclear technology, such as research and medicine. But for hybrids, radioactive waste from fusion is restricted to neutron activated structural materials.

Radioactivity   The process of unstable atomic nuclei that spontaneously decompose to form nuclei with a higher stability. The energy and particles which are released during the decomposition process are called radiation: alpha particles are helium nuclei, beta rays are electrons, gamma rays are electromagnetic radiation. The process occurs in nature: natural radioactivity.

Rocket effect   The effect of dense matter being propelled in the direction opposite to the high velocity motion of ablated material. It is a consequence of momentum conservation.

Rotational transform (or field line pitch)   On a toroidal magnetic surface, successive intersections of a field line by a given meridian plane.

Safety factor   Number of toroidal roundtrips divided by the number of poloidal roundtrips of a field line on a toroidal magnetic surface.

Scaling law   Theoretical or semi-empirical relationship between major plasma variables and the size of a confinement device.

Shear   Property associated with the rotation of magnetic field lines along a perpendicular path. A dipole field has no shear, contrary to the magnetic field map of a tokamak.

Shock wave, shock front   When a high pressure is suddenly applied to some material, matter is compressed, heated and set into motion. A pressure discontinuity called a shock wave or a shock front propagates through the material.

Sub-critical   Assembly of fissile elements in which the effective multiplication coefficient is smaller than 1. Any neutron population here can only decay.

Tokamak de Fontenay aux Roses (TFR)   One of the first tokamaks built in western countries. A very successful first generation device.

Thermal energy, thermal equilibrium   Energy associated with disordered motions and oscillations of an assembly of elementary objects. Thermal equilibrium is obtained when heat exchanges with the external world vanish, then the temperature is related to the average energy via a simple formula.

Thermonuclear gain (inertial fusion)   Ratio between fusion energy and the energy of the driver: laser beam or any other source that serves for implosion and ignition of the thermonuclear fuel.

Thermonuclear gain (magnetic fusion)   Ratio between fusion power and the power of auxiliary heating that maintains a high plasmas temperature.

TOKAMAK (in russian: TOroïdalnaia KAmera i MAgnetnaia Katouchka, for toroidal chamber and magnetic coil)   High-intensity current-carrying plasma ring. In the late 1960s, a breakthrough in confinement performances was obtained with such a device.

Triple product   In a magnetically confined plasma, product of the ion density, the confinement time and the temperature. Its value is a figure of merit for magnetic confinement. Fusion conditions correspond to a triple product exceeding 3×10^{22} cm^{-3}sK.

Tunnelling   Due to its wave nature, a charged particle with kinetic energy below the potential barrier has a crossing probability that can be calculated thanks to laws that are similar to the laws of optics.

Acronyms

ADS	Accelerator Driven Systems,
BNL	Brookhaven National Laboratory
CEA	Commissariat à l'Energie Atomique et aux Energies Alternatives (France)
CPRP	Contemporary Physics Education Project. Chapters dealing with fusion come from PPPL.
EPR	European Pressurized Reactot
FLIBE	Fluor Lithium Béryllium ($2LiF + BeF_2$)
HIPER	HIgh Power laser Energy Research facility
FWHM	Full Width Half Maximum
IEC	Inertial Electrostatic Confinement
IFMIF	International Fusion Material Irradiation Facility
ITER	International Thermonuclear Experimental Reactor
LBNL	Lawrence Berkeley National Laboratory
LLNL	Lawrence Livermore National Laboratory
LIFE	Laser Inertial Fusion-fission Energy
LMJ	Laser MégaJoule (France)
MAST	Mega Ampere Spherical Tokamak
NIF	National Ignition Facility

NST	National Spherical Torus
ODS	Oxide Dispersion Strengthened
PETAL	PETawatt Aquitaine Laser
PPPL	Princeton Plasma Physics Laboratory
PWR	Pressurized Water Reactor
RAFM	Reduced Activation Ferritic Martensitic
TFR	Tokamak de Fontenay aux Roses (France)

Selected Bibliography

1. *Proceedings of the Second United Nations Conference on Peaceful Uses of Nuclear Energy, Vols. 31 & 32*, United Nations, Geneva (1958).
2. *Controlled Thermonuclear Reactions*, S. Glasstone & R. H. Lovberg, Van Norstrand (1960).
3. *Plasmas and Controlled Fusion*, D. J. Rose & M. Clark, MIT Press (1961).
4. *Controlled Thermonuclear Reactions*, L. A. Artsimovitch, Gordon and Breach (1974).
5. *Fusion Reactor Physics*, T. Kammash, Ann Arbor Science (1975).
6. *The Physics of Laser Fusion*, H. Motz, Academic (1979).
7. *Plasma Physics and Nuclear Fusion Research*, R. D. Gill (ed.), Academic (1981).
8. *Fusion*, J. L. Bromberg, MIT Press (1982).
9. *Fusion Energy*, R. A. Gross, Wiley (1984).
10. *Fusion: The Search for an Endless Energy*, R. Herman, Cambridge University Press (1990).
11. *The Fusion Quest*, T. K. Fowler, John Hopkins University Press (1997).
12. *Fusion: A Voyage Through the Plasma Universe*, H. Wilhelmsson, I.O.P. (1999).
13. *High Power Laser Interactions, Isotopes Separation, Nuclear Fusion Control, Elementary Particles Selective Creation*, J. Robieux, Tech. & Doc./Lavoisier (2000).
14. *Principles of Fusion Energy: An Introduction to Fusion Energy for Students of Science and Engineering*, A. A. Harms, K. F. Schoepf, G. H. Miley, D. R. Kingdon, World Scientific (2000).
15. *Nuclear Fusion. Half a Century of Magnetic Confinement Fusion Research*, C. M. Braams & P. E. Stott, Taylor & Francis (2002).
16. *Tokamaks*, 3rd ed., J. Wesson, Clarendon Press Oxford, (2004).
17. *Plasma Physics and Controlled Nuclear Fusion*, K. Miyamoto, Springer (2005).
18. *Plasma Physics and Fusion Energy*, J. P. Freidberg, Cambridge University Press (2007).

19. *Energy for the future, the nuclear option*. European Physical Society position paper (2007) http://www.eps.org/highlights/energy-for-the-future/.
20. *The Physics of Inertial Fusion: Beam Plasma Interaction, Hydrodynamics, Hot Dense Matter*, S. Atzeni & J. Meyer-ter-Vehn, Oxford (2009).
21. *Fusion, An Introduction to the Physics and Technology of Magnetic Confinement Fusion*, 2nd ed., W. M. Stacey, Wiley (2010).
22. *An Indispensable Truth: How Fusion Power Can Save the Planet*, F. F. Chen, Springer (2011).

A few books in French:

23. *L'énergie thermonucléaire*, C. Etiévant, P.U.F. Que sais-je? N° 1017, 2ème édition (1976)
24. *L'énergie des étoiles*, P. H. Rebut, Odile Jacob (1999)
25. *Les déconvenues de Prométhée, la longue marche vers l'énergie thermonucléaire*, J. L. Bobin, Atlantica (2001)
26. *ITER, le chemin des étoiles*, R. Arnoux & J. Jacquinot, Edisud (2006)
27. *L'énergie bleue*, G. Laval, Odile Jacob (2007)
28. *La fusion nucléaire*, A. Benuzzi-Mounaix, Belin (2008)
29. *La fusion nucléaire par laser*, J. Robieux, éditions Louis de Broglie (2008)
30. *La fusion nucléaire: de la recherche fondamentale à la production d'énergie?* Rapport de l'Académie des Sciences sous la direction de Guy Laval, EDP-Sciences (2007)

Index

microballoon, 109–112, 114, 132, 134–136, 138, 140, 141, 187
Murphy's law, 46, 187

negative ions, 70
neoclassical transport, 55
neodymium glass (laser), 25, 102, 106, 121, 127, 129, 136, 149
NIF, xiii, xiv, 112, 121, 126, 127, 130–132, 134, 149, 168, 173, 174, 180, 187, 191
nonlinear effects, 39, 40, 103, 108
NST, 187, 192

Oklo, 6
oscillator (laser), 121, 123, 124, 126, 134
overdense zone, 102

parallel amplification, 121
parametric instabilities, 104, 118, 129
passing trajectories, 53, 54
Pfirsch and Schlütter (regime), 55
Plank's formula, 59
Poincaré, H., 35
poloidal field, 51, 62
potential barrier, 9–12, 187, 189
pumping (laser), 120, 123–125, 129

radioactive waste, 146, 149, 170, 180, 181, 187
radioactivity, 4, 5, 68, 74, 86, 87, 110, 111, 170, 187
Rayleigh–Taylor instability, 105–107, 133
reaction rate, 15–18, 105, 154–156
Rebut, P. H., 73
resonance (magnetised plasmas), 13, 29, 37, 38, 55, 56, 66, 104, 155, 157, 184, 186
Richtmyer–Meshkov instability, 105
rocket effect, 102, 187
rotational transform, 49, 50, 58, 188
RTL, 143, 151
Rutherford, E., xi, 1, 2, 5, 15

safety, 24, 86, 146, 160
safety factor, 50, 51, 55, 63, 64, 67, 74, 118, 188

Sakharov, A., 49, 146
scaling law, 63–65, 72, 80, 81, 188
Schmitt, H. H., 178
shear, 50, 54–56, 67, 74, 188
shock wave, 99, 101, 102, 105, 134, 151, 188
Spitzer, L., 58
subcritical, 145, 146, 149, 183, 188

TFR, 65, 70–72, 188, 192
TFTR, 68, 71, 73–75, 79
thermal barrier (inertial fusion), 31, 32, 102, 104, 109, 111
thermal barrier (magnetic fusion), 31, 32, 102, 104, 109, 111
thermal energy, 15, 18, 21, 24, 29, 68, 99, 104, 147, 188
thermodynamic trajectory, 100, 101
thermonuclear gain (inertial), 100, 117, 119, 140, 147–149, 163, 167, 168, 188
thermonuclear gain (magnetic), 64, 74–76, 83, 163–166, 188
Tokamak, xiii, xiv, 38, 49–54, 56–58, 61–63, 65–68, 70–77, 79–84, 86, 87, 91, 93–95, 145, 147, 148, 158–161, 164–167, 169, 171–173, 175, 179, 180, 185, 186, 188, 191, 192
toroidal field, 48, 49, 57, 58, 62, 66, 82, 148
transport (coefficients), 31, 55, 103
trapped trajectories, 54
triple product, 64, 65, 74–77, 166, 179, 189
tritium, xiv, 9, 12–14, 23–25, 68, 74, 75, 80, 83, 85, 86, 91, 96, 111, 146, 149, 154, 155, 162, 171–173, 176, 177, 183
tunnelling, 10, 11, 13, 189
turbulence, 28, 36, 38–40, 54, 72, 123

under-dense zone, 102, 103

von Weizsäcker, C. W., 11

wave coupling, 39, 61, 62, 104, 140
Wendelstein 7-X, 59, 173

YAG, 123